39 Advances in Polymer Science

Fortschritte der Hochpolymeren-Forschung

W0044006

Editors: H.-J. Cantow, Freiburg i. Br. · G. Dall'Asta, Colleferro · K. Dušek,
Prague · J. D. Ferry, Madison · H. Fujita, Osaka · M. Gordon, Colchester
J. P. Kennedy, Akron · W. Kern, Mainz · S. Okamura, Kyoto
C. G. Overberger, Ann Arbor · T. Saegusa, Kyoto · G. V. Schulz, Mainz
W. P. Slichter, Murray Hill · J. K. Stille, Fort Collins

Polymer Products

With Contributions by
L. Andreeva, K. Geckeler, M. Mutter, V. N. R. Pillai,
F. A. Shutov and V. Tsvetkov

With 111 Figures

Springer-Verlag
Berlin Heidelberg GmbH 1981

ISBN 978-3-662-15386-4   ISBN 978-3-540-38282-9 (eBook)
DOI 10.1007/978-3-540-38282-9

Library of Congress Catalog Card Number 61-642

Table of Contents

Foamed Polymers Based on Reactive Oligomers

Fyodor A. Shutov

Physical Department, Building Institute, Leningrad 198005, USSR

To the Memory of my Friend and Teacher Professor Alfred A. Berlin

Daß ich erkenne, was die Welt
im Innersten zusammenhält,
schau' alle Wirkungskraft und Samen
und tu' nicht mehr in Worten kramen.

J. W. Goethe

This article surveys the specific features of polymer science for gas-filled polymers (main principles of their classification as well as the major problems of their physics, chemistry, technology and application). Special attention is given to the discussion of the author's findings concerning the morphology of foamed plastics based on reactive oligomers such as polyurethane, phenolic and urea-formaldehyde polymers. These findings evidence the presence in their cellular structure of micromorphological cells (size: 0.01–0.1 micron), which the author terms microcells. *It is especially emphasized that these microcells represent the most fundamental and most widely used type of such foamed plastics. The number of these microcells is $10^2 - 10^3$ times higher than that of the well-known* macrocells; *this is why the specific surface area of such polymer substances is larger than 200 m^2/g. The physicochemical properties of oligomeric foamed polymers (thermooxidation, electrical and moisture absorption) are explained by the microcell concept. Future trends with regard to new starting materials, methods of preparation, technology, theoretical investigation and long-term perspectives of these plastic foames are discussed.*

Table of Contents

Abbreviations

BET	Brunauer-Emmett-Teller method	PEN-1	Soviet grade of epoxide-novolac phenolic foam
DABCO	triethylenediamine/dipropylene glycol	PF foam	phenol-formaldehyde foam
DC	polymethylsiloxanes	PPU-3,	$-3\,\text{S}, -102, -305, -305\,\text{A},$
FL-1	Soviet grade of resol phenol-formaldehyde foam		-307 Soviet grades of PUR foams
		PUR foam	polyurethane foam
FRP-1	Soviet grade of resol phenol-formaldehyde foam	RO	reactive oligomer
GSE	Gas-Structural Element	SE	static electrization
MDI	4,4'-diphenylisocyanate	SEP-1	Soviet grade of phenolic-urethane foam
MGF-1	a, ω-methacryl(bisethylene-glycole) phtalate	SFUP	Soviet grade of phenolic-urethane foam
MGF-9	a, ω-methacryl(diethylene-glycole) phtalate	SIN	Simultaneous Interpenetrating Networks
NMR	nuclear magnetic resonance method	TGM-3	tris(oxyethylene)-a, ω-(dimeth) acrylate
OEA	oligoester acrylate	TMGF-11	a, ω-dimethacryl (1,3) bis-glycerine-2-phtalate
OEM	oligoester methacrylate		
OFM	oligoester fumarate maleates	UF foam	urea-formaldehyde foam

Physicochemical Symbols

a	sample thickness; coefficient	ΔP_f	drop of the pressure of liquid
A	surface area of foamed plastic	ΔP_g	drop of the pressure of water steam
A_0^*	amplitude of free nuclear induction	ΔP_c	suction pressure of capillaries
C_1, C_2, C_3	empirical coefficients	q	electrostatic charge
d	thickness of struts; diameter of the water molecule	Q	heat flow
		r, R	cell radius
d_f	hydraulic diameter	R_1, R_2	curvature radius
D	diameter of cell; the most probable size of cells	Re	Reynolds number
		S	specific surface area
D_i	current diameter of cells	T	temperature
E_{br}	breakdown voltage (dielectric strength)	v	flow rate
f	frequency of electromagnetic field	V	geometric volume of foam
$F(x)$	Weibull function	W	weight humidity
h	Planck constant	u	contact angle
H	molar heat of phase transition	γ	apparent density of foam
k	Boltzmann constant	γ_f	density of liquid
K	empirical coefficient	γ_g	density of gas
K_1, K_2	coefficients of uniformity of cellular structure	γ'_g	saturation density of gase phase
		γ_p	density of polymer phase
l	length of rib of dodecahedron	γ_w	density of water
l_a, l_b, l_c	traces of current via liquid foam	γ_1	gyromagnetic ratio of the nucleus
L	capillar length	Γ	gamma-function
m	mass of dry foam	δ	thickness of cell wall
m_g	mass of gas	tg δ	dielectric losses
m_w	mass of water	ϵ	dielectric permeability
M	the arithmetic mean value	η_f	kinematic viscosity
n	number of rows of foam models	κ	electroconductivity
P	porosity; pressure	ϑ	volume fraction of phase
P_0	saturated vapor pressure	ϑ_g	volume fraction of gas

ϑ_p volume fraction of polymer phase

ϑ_w volume fraction of water

ϑ_α volume fraction of open cells

ϑ_w^* volume fraction of equilibrium water

ϑ_w^1 volume fraction of the monomolecular layer of absorbed water

ϑ_{max}^* maximal value of ϑ_w^*

Θ correction factor

λ generalized conductivity

λ_i D_i/D

ν_i the frequency of appearance of cells with size "i"

ρ_v volumetric electroresistance

σ surface tension coefficient; surface density of charge; root-mean-square deviation

τ current time

τ^* time of equilibrium moisture absorption

τ_1 longitudinal relaxation time

φ relative humidity

χ Pirson criterion

ψ resistance coefficient

ω volume humidity

1 Introduction

In polymer science and technology there is a discrepancy between the ever-increasing demand for polymer materials with improved properties and the existing energy and raw material resources of chemical technology. One of the basic solutions to this problem is the development of chemico-technological and physico-technical principles which would enable to obtain materials with the required properties from liquid (or readily fusible) inexpensive organic compounds, which can be converted to polymeric articles directly during processing or synthesis, without additional supply of heat and pressure.

This strategy may be realized by the use of reactive oligomers (RO), i.e. low-molecular weight compounds which may be converted to polymers of linear, branched, ladder and three-dimensional network structures. Of special importance are RO's which form cross-linked polymers since in this case materials with optimal values of heat and fire resistance, strength, chemical stability, atmospheric resistance, durability, etc. may be obtained.

Thus, the use of RO provides a novel route in polymer technology based on the direct conversion of "liquid (RO) — articles", avoiding the lengthy, complex and energy-consuming stages of isolation, processing and manufacture of polymer articles from high-molecular compounds. With RO it is possible to combine the processes of formation of the polymer itself with those of the article in a single process, i.e. to realize "reactive" or "chemical" molding of liquid (or readily fusible) compounds at ambient (or mild) temperatures, following the terminology proposed by A. Berlin[1, 2].

Problems related to oligomeric foams are only part of the more general problems arising in the field of polymer science dealing with gas-filled polymers. In the author's opinion, the specificity of scientific and applied problems of oligomeric foams may be formulated accurately and clearly only by considering them from the point of view of this science as a whole, i.e. using the deductive method. Thus, this survey begins with a consideration of the general principles of classification and discussion of scientific and applied problems, common to all types of gas-filled polymers.

On the other hand, for the understanding of the importance of the achievements made in solving specific problems (in our case relating to oligomeric foams) and in the advance of polymer science as a whole, one should use the inductive method. Therefore, this survey is concluded by a discussion of the trends and perspectives of development of all gas-filled polymers.

Only isotropic gas-filled polymers based on RO will be considered. Anisotropic (integral) and syntactic foams based on RO are discussed in[3]; general problems of the physical chemistry and technology of polymer foams based on high polymers and reactive oligomers are dealt with in[4, 5].

2 Principles of Polymer Foam Classification

Mankind has long been interested in natural gas-filled materials. The combination of lightness and buoyancy with strength and good heat insulation properties ensured the wide use of the most common of porous materials, wood. Other natural porous materials such as leather, felt, asbestos, pumice are also of exceptional practical importance. Not so long ago several man-made materials were added to this list, for example, porous concrete, foam ceramics, foam glass, foamed metals.

The development of technology, especially aviation and the building industry, created more stringent requirements which could not be met by existing natural and man-made materials. Research began in the late 1930's and early 1940's in many countries with the aim of creating new gas-filled materials based on synthetic organic compounds.

At present, hundreds of various elastic and rigid gas-filled materials used literally in all branches of industry are produced on the basis of reactive oligomers and high polymers. The production of these materials is rapidly expanding. Thus, in 1970 the world output of plastic foams was 2 million tons, in 1975 3.5 million tons, 1980 it will be 5−6 million tons, and in 1985 about 20% (6% in 1975) of all plastic materials will be gas-filled[5].

2.1 Three Generations of Foamed Polymers

Increase of the output of gas-filled plastics is connected first of all with the development of new types of compositions and of new highly efficient production methods. These materials are produced both under industrial conditions on transfer lines and directly at locations by casting or spraying.

In the last decade several novel types of gas-filled polymers appeared which belong to the second generation: integral (structural) plastic foams; syntactic foams; reinforced polymer foams; multilayer foams (foamed laminates), metallized plastic foams; mineral and metallic foamed materials obtained on the basis of foamed polymers; laminated constructions on the basis of foamed polymers and monolithic (unfoamed) plastics, metals, paper, leather, etc.

Manufacture of second generation materials requires not only different compositions, technological conditions and equipment but also, although to a lesser extent, completely new technological approaches and physicochemical principles. Thus, in the preparation of integral foamed polymers it is necessary to solve a problem that is inverse to the one that existed and exists in the technology of foamed plastics. Indeed, a "good macrostructure" was previously considered to be one with uniform distribution of specific gravity over the entire volume of the article, and the ratio of components, conditions of foaming, etc. were selected so as to ensure this uniformity. On the other hand, a good structure in integral foams is one with a completely different physical pattern, non-uniformity of gas filling and of foamed polymer density over the volume of the article (the density increases from the center to the periphery) and the higher this non-uniformity the higher is the quality of the foamed polymer and the better are its properties[3].

Gas-filled polymers of the third generation, which are emerging now, are mainly the "children" of second-generation plastic foams: filled, reinforced, "one-sided" and "inverse" integral foams; syntactic-integral foams; anisotropic foamed polymers; foamed films and foamed fibers (including integral ones). Impregnated plastic foams, including those impregnated with liquid foam compositions (materials of the type "plastic foam in plastic foam") also belong to the third generation. These materials

are produced by even more sophisticated and costly technology than the second
generation materials.

2.2 Gas Filling and Properties of Foamed Plastics

The special position of gas-filled plastics among polymeric composites is primarily
due to the extraordinary combination of lightness with relatively high strength and
very high heat, sound and electrical insulation properties. This combination of properties is determined by the peculiar macrostructure of these materials: the polymeric
matrix is not monolithic but contains discontinuities due to gas inclusion distributed
over the entire bulk of the substance. In the limiting cases the gas may be the discon-
tinuous phase (as in superheavy plastic foams or gasified plastics) or the continuous
phase (as in superlight foamed plastics).

Gas-filled polymers may formally be considered as filled polymeric compositions
where air or another gas is used as a filler. The characteristics of such fillers are very
unusual. Thus, density and tensile strength are by several orders of magnitude lower
than those of the polymeric matrix. Depending on the amount of filler, the volume ratio
of gaseous and polymeric phases may vary from 0.1:1 to 30:1. In the latter case the
system may be considered not as a gas-filled, but as a "gaseous" composition, where
the walls and struts of polymer cells act as reinforcing elements. The viability of such
an approach is confirmed for example by the reliability of the formulae for calculating coefficients of heat- and temperature conductivity for several superlight foamed
polymers.

The introduction of a considerable volume of gaseous phase into the polymer
causes a marked change in physical properties: volume weight (apparent density);
permeability of liquids and gases; capacity to redistribute local mechanical stresses;
buoyancy; etc. Furthermore, depending on the conditions of formation, degree of
gas filling and the type of morphological structure, properties of the polymeric matrix
itself practically always change also (orientational effects of the supramolecular structure of cell walls and cell ribs or struts), resulting in a pronounced increase of relative
strength in relation to impact loads, in anisotropy of elastic properties, etc.

The high degree of gas filling "competes" with the specific features of the chemical structure of the polymeric matrix, levelling the properties of various foamed materials. Thus, for large gas inclusions (superlight and light foamed polymers) several
important technical characteristics of nearly all plastic foams (coefficient of thermal
conductivity, dielectric permeability, tangent of dielectric loss angle) are practically
equal and are independent of the chemical structure of the initial polymer (see
Chap. 7.1). Therefore, gas filling of polymers results in smoothing out of differences
in physical properties.

2.3 Gas-Structure Element as the Main Morphological Unit

It is now apparent that the classification of porous materials which was based on the
cell (pore) and type of communication between cells (pores), as the main morphologi-

cal unit requires further improvement. Analysis of the results obtained in the last few years shows that the main morphological unit which determines the properties of gas-filled polymers is not only the cell itself (its shape and size), but also the size and configuration of the intercellular space filled with the polymeric matrix, i.e. cell walls and ribs. Indeed, it has been established for various types of foamed polymers that for the same size and shape of cells and fraction of communicating cells, the physico-mechanical properties of gas-filled plastics, for example strength and elasticity, may differ considerably on account of the differences in the shape and sizes of ribs or in wall thickness.

In accordance with the above-mentioned, it is necessary to introduce a new morphological notion, "a unit of morphology", a gas-structure element (GSE), replacing the traditional "cell". Hence, the term GSE stands for a primary spatial structure consisting of a gaseous cavity (cell), its walls and ribs, i.e. the elementary volume of gaseous and solid phases which is repeated with certain periodicity and with a high degree of regularity in the entire material and makes up the macrostructure of the foamed polymer. The GSE concept is not qualitative but quantitative, since GSE is characterized by the shape and size, type of packing and the distribution in the foam bulk. Obviously, the GSE concept is more general than the "cell" concept: cells with the same size and shape may form different types of gas-structure elements due to the different configuration and structure of the intercellular space. On the other hand, the equality of GSE involves equality of the shape and sizes of the cells and also of the walls, struts and intercellular space[4, 5].

2.4 New Morphological Classification

On the basis of the GSE concept gas-filled materials may be divided into the following types:
1. Cellular (or foamed) materials containing isolated GSE;
2. porous materials containing interconnected GSE;
3. microballon or syntactic foams, the GSE of which consists of a gaseous phase, included in a spherical shell of monolithic material, and a solid phase composed of these shells (of polymeric or mineral nature) and of the interspherical space filled with polymer;
4. honeycomb plastics, the GSE of which consists of a gaseous phase, included in regular polyhedra, and a solid phase as in the previous case composed of two materials, e.g. polyhedra faces of paper or fabric, and a polymeric binder;
5. capillary of fibrous plastics, whose GSE consist of a gaseous phase included in an anisometrically shaped volume and of a polymeric binder;
6. foamed materials with a mixed type of GSE.

This classification of foamed materials only formally resembles the classification based on the "cell" concept proposed by A. A. Berlin more than a quarter of a century ago[6]; it is based on quite different morphological notions about the spatial structure of foamed polymers.

The limiting case of porous plastics are the so-called reticular porous plastics which lack cell walls, and the entire polymer phase is concentrated in the cell struts;

such materials may also be called "absolute porous plastics". It will be noted that the morphology of reticular foams cannot be described by using the "cell" concept since such materials do not contain cells in the proper sense of the word, whereas GSE's and their characteristics may be determined both qualitatively and quantitatively[5].

While reticular porous plastics are the limiting case of the principle of formation of gas-filled plastics with interconnected GSE, the so-called syntactic foams, by analogy, are "absolute foamed plastics" since all GSE's in these materials are isolated from each other. The latter may also be called "physical foams" since the cellular structure of these materials is formed not as a result of a complex combination of colloidal and chemical phenomena, which accompany the process of foaming, but by filling monolithic compositions with microspheres (microballons). These contain air or other gases due to physical (mechanical) introduction of fillers, excluding all physico-chemical interactions between gas and polymeric matrix during the preparation of the foam[3].

Also, the preparation of honeycomb plastics and gas-filled fibrous materials does not require foaming. Formation of gas cavities in the former is carried out by glueing of previously prepared sheet materials or by extrusion of thermoplastics through special nozzles; in the latter case by glueing or filling with polymeric binders of fibrous substances containing capillaries and pores.

2.5 Classification According to Physicochemical Principles of Preparation

In addition to the GSE classification of foamed polymers, which may presumably be applied not only to organic gas-filled materials, it is also necessary to distinguish gas-filled polymers according to the method of preparation. When carrying out such systematization it should be taken into account that foamed plastics are now prepared from practically all types of synthetic materials, i.e. from high molecular weight compounds, polyreactive oligomers and from polymer-oligomer (monomer) compositions.

In the preparation of gas-filled plastics from high polymers or their mixtures, the same technological procedures and equipment are used as for the preparation of the corresponding monolithic plastics (pressing, extrusion injection, molding, sintering, etc.) and the foaming process itself is carried out at temperatures either close to or exceeding the temperature of viscous flow of the polymers[5].

In contrast, oligomer technology is based on the conversion of liquid oligomers into foams by direct transition of the liquid into the polymer (mainly of network structure). Such "chemical" formation does not require high temperatures and pressures and hermeticity of the molds, thus greatly reducing expenditure on equipment, energy and labor. For example, the use of oligomeric compositions made it possible to prepare foamed plastics *in situ* by casting and spraying[4].

Finally, in the preparation of gas-filled materials on the basis of polymer-oligomer compositions, oligomers which are plasticizers or lubricants of the corresponding high polymers are used. Such compositions ensure miscibility of components with blowing agents, fillers, stabilizers, etc. and subsequent foaming of polymers is usually carried out at lower temperatures than in the absence of oligomers. This technique has long been used for the industrial production of several grades of plastic foams on the basis of poly(vinyl chloride) and polystyrene.

3 Problems of Polymer Science

3.1 Empirical and Scientific Methods of Investigation

Knowledge of the chemical and physical laws governing the formation of polymeric foams has always lagged behind the achievements of technology. Even today, in most cases, we cannot theoretically explain why, under given conditions, a material possesses certain physicochemical characteristics. Achievements of technologists are mainly connected with a purely empirical approach to the preparation of new grades and designs of new processes, although such an approach is not very efficient. Indeed, a mixture for preparation of a gas-filled polymer may contain up to ten or more components and, evidently, examination by the trial and error procedure of hundreds of versions of recipes and concentrations using even a computer requires enormous expenditure of labor and materials.

However, the history of the science of construction materials, including polymeric ones, reveals that the solution to the main problem, namely the creation of high-quality materials with the required properties, may be achieved only when using a strictly scientific approach.

This somewhat critical situation may be resolved by the determination of specific and general physical and chemical regularities governing the formation and behavior of polymeric foams, which requires the use of a wide range of ideas and techniques developed in other sciences: physical and colloidal chemistry, physicochemical mechanics, rheology, thermodynamics, physics of polymers, physics and mechanics of non-continuous media, physics of surface and transfer phenomena, chemical physics of oxidation and degradation processes, etc.

3.2 A Common Physicochemical Approach and Specificity of Scientific Problems

During the last 25 years many surveys, monographs and other works have been published on general and specific problems of the chemistry, technology, properties and applications of foamed polymers. Several works give a survey of specific problems of foamed plastics, but even these works lack a consistent physicochemical approach. As a result, the empirical approach stands firm. Although a physicochemical concept of foam formation for the case of low molecular weight compounds has been developed and provides very useful data, it is applicable only to the initial stages of foaming of polymer solutions and melts.

In the author's opinion, the analysis of the problems of chemistry, technology and properties of plastic foams should be carried out both from the viewpoint of the chemistry and physics of condensed systems, including the physics of polymers, and from the positions of physics and chemistry of dispersed systems, particularly heterogeneous gas-solid systems (see Chap. 5.4).

A common approach to the study of properties, in particular variations of properties under external effects, should include, as one of the basic ideas, the conception of polymeric foams as heterogeneous systems with developed surface (see Chap. 5.4). As became apparent in the early 1970's, the specific surface of these materials is of the order of dozens and even hundreds of square meters per gram[4, 7-11]. Such high values of the specific surface are quite exceptional among polymeric materials and lead to marked (in relation to expected) changes in several physicochemical properties of plastic foams, for example water absorption and resistance to thermal oxidation. As a result, the service temperatures for foams of the same grade but of different dispersity may differ by several dozens of degrees, while heat conductivity and electrophysical properties may differ by one order of magnitude[7].

The solution of the basic problem of the science of construction materials, namely the establishment of the relationship between structure and properties acquires special complexity for gas-filled polymers. Six levels of structural organization can be distinguished in these materials:

1. Chemical composition or primary structure of the initial polymer;
2. secondary structure-conformation of linear or reticulate polymer molecules;
3. supermolecular structure of cell walls and struts;
4. macrostructure (GSE type);
5. microcellular structure of GSE — microcells on the surface and in walls and struts of macrocells;
6. supercellular structure — GSE distribution in the bulk, distribution of volume weight over the width and height of foam products.

For practical calculations of durability and for the determination of application conditions it should be taken into account that these materials are composed of a three-phase structure (gas — solid — liquid). A certain amount of a liquid phase, due to the condensation of water vapor in the air, is practically always present inside plastic foams. The presence of the liquid phase plays a decisive role in mass, gas and heat transfer and sharply reduces the heat and electrical insulation properties of polymeric foams (see Chap. 6).

The rapidly increasing use of foamed plastics has stimulated the study of their properties and of the relationship between technical properties, composition, method of preparation and the effect of environmental factors. Nevertheless, attention to the problems of ageing as a result of thermal, thermal oxidative and radiation degradation is insufficient.

A more profound understanding of the behavior of plastic foams under various exploitation conditions will make it possible to estimate service conditions and also to determine the precise and most rational functional application of a given foamed polymer. Unfortunately, some foamed materials are still erroneously rejected altogether, only because their properties once did not fit specific service conditions.

Today, the research of construction materials involves transition from the phenomenological to the physicochemical approach, i.e. to the study of the molecular mechanisms of degradation processes using methods of the physics and physicochemistry of polymers. The practical result of these investigations will be scientifically substantiated rather than an empirical selection of effective inhibitors of degradation processes. The solution of these problems is especially important for polymer foams since, as gas-filled systems with a well developed surface, they are particularly vulnerable to thermal oxidation due to the presence of an aqueous oxidizing agent — atmospheric oxygen. It should be noted that the selection of inhibitors for the thermal oxidation of foams is an independent problem which cannot be solved "by analogy", knowing the type and concentration of the inhibitor effective for the corresponding monolithic polymer. The results obtained in the last few years unambiguously indicate that many inhibitors of non-foamed plastics, when introduced into the polymeric matrix of foams, enhance rather than inhibit thermal oxidation[4, 7].

3.3 Classification of Theoretical and Applied Studies

Summing up the details mentioned above, let us outline the main trends in the science of plastic foams:

Chemical: search for new and improvement of known recipes and compositions, increase of the number of different foam-forming oligomers and polymers, blowing agents, surfactants, cross-linking agents, etc.

Technological: development of new methods of foaming, design of highly effective equipment for the forming and foaming of compositions and for secondary processing of materials.

Physicochemical: studies of the regularities of foaming and formation of cellular structure in foams, studies of rheological and thermodynamic parameters of foaming and of the mechanism of interactions between components, etc.

Construction-material: studies on the influence of the chemical structure and specific features of the macro- and microstructure on the physicomechanical properties of foamed polymers, etc.

Physical: studies of molecular mechanisms, physical and chemical phenomena and elucidation of changes in foam properties under the effect of external factors; studies on the chemical structure and morphology using methods of spectroscopy, X-ray analysis, microscopy, thermography, etc.

Applications: determination of durability, service conditions, development of methods for calculating technical characteristics of foams that take into account the effect of external factors, extension of the range of application of foams, etc.

These routes are at different stages of development. Due to the popularity of the empirical approach, the chemical and technological directions are better developed than the others. Technical characteristics of foamed plastics, which are in many respects unique, determine the rapid development of the applied direction. Physicochemical, physical and construction-material directions are lagging far behind. However, these are the routes which ultimately provide the scientific foundations for gas-filled polymer production.

4 Promising Oligomeric Foams

The ever increasing demand for foamed polymers stimulates research aimed at the extension of the raw material base and variety of these materials. In this connection it is necessary to consider the types of oligomeric foams which have not yet "left the laboratories", but which, in our opinion, are very promising for the further development of the industry of gas-filled polymers.

Without considering in detail the chemical classification of RO, we shall only note that all types of RO may be divided into two groups[1, 12]:

1. RO capable of hardening without formation of side products – polymerizable oligomers.
2. RO which form polymers with the formation of low-molecular weight products – polyreactive oligomers.

The first group includes oligoesteracrylates, oligoestermaleates, oligoester- and oligodieneurethane-isocyanates, oligomeric epoxides, polymerizable oligosiloxanes, oligodienes ("liquid resins"), oligoimides, oligobenzimidopyrrolidones, etc.

The second group includes phenol-formaldehyde and urea-formaldehyde (carbamide) oligomers, di- and polymercaptooligosulfides, oligosiloxanedioles, carboxylic group-containing oligomers with di- and polyisocyanates, etc. The majority of commercial grades of oligomeric foams are prepared from polyreactive oligomers.

4.1 Unsaturated Oligoesters and Foamed Plastics Therefrom

Plastic foams obtained from oligoesters capable of hardening with the formation of network polymers or copolymers are becoming increasingly important. Two classes of polymerizable oligomers are most promising in this respect, namely oligoesters based on fumarates and maleates (OFM) and oligoesters based on methacrylates (OEM) and their analogs. The formers are obtained by oligocondensation of maleic anhydride, dibasic acids (or their anhydrides), and glycols (or α-oxides)[1, 2]. The process is usually carried out at $180-220\,^{\circ}C$, i.e. when more than 80% of maleates isomerize to the more reactive fumarates:

$$HC = CH \atop OC \diagdown_{O}\diagup CO + R(OH)_2 + R'(COOH)_2 \xrightarrow{-n\,H_2O} -O\text{-}[COCH=HCCOORO]\text{-}[COR'COORO]\text{-}$$

where R and R' are the hydrocarbon residues of the diol and dicarboxylic acid, respectively.

OFM are viscous liquids, or readily fusible resin-like substances, capable (in the presence of intiators or upon irradiation) of three-dimensional radical copolymerization with several active monomers (styrene, vinyltoluene, methylmethacrylate and others). The properties of the network copolymers formed largely depend on the chemical nature of the reagents and the copolymerization conditions. Hardening of OFM may be carried out at room temperature, or even lower, using mixtures of peroxide initiators: cumene hydroperoxide, methyl ethyl ketone peroxide (MEKP), benzoyl peroxide, etc. and activators which reduce the activation energy of the radical decomposition of initiators. Organic bases, oxides and transition metal (manganese, cobalt, tungsten, and others) salts are usually used as activators. Low-temperature hardening of OFM may also by performed by photopolymerization (near ultraviolet) or radiation polymerization (γ-rays, electrons).

The network polymers formed after hardening exhibit high strength and good electrical insulation properties, adhesion to polar surfaces and prolonged heat resistance at 150–180 °C. In order to improve fire resistance of OFM, residues of chlorinated dibasic acids and (or) diols with (or without) antimony oxide or phosphorous-containing comonomers are added.

However, OFM have several disadvantages: they do not undergo three-dimensional homopolymerization, reveal limited miscibility with some oligomers and high polymers and a statistical distribution of reactive groups and chain units. In order to eliminate these shortcomings, (highly toxic) monomers which increase shrinkage and internal stresses are used during hardening. The statistical distribution of reactive groups and chain units in OFM molecules and the limited miscibility of the latter impair the modification of the composite properties of OFM and restrict the use of these oligoesters. Nevertheless, OFM are produced on a large scale and they find application as binders for glass reinforced plastics, as bases for protective coatings, electrical insulating compounds, adhesives and sealings[2, 13].

The other class of unsaturated oligoesters – oligoester methacrylates (OEM) and oligoester acrylates (OEA) and their numerous analogs – were first synthesized by A. Berlin and co-workers[6] in the 1940's and industrially utilized in the early 1950's. In contrast to OFM, OEM and its analogs contain regularly alternating reactive acrylic groups which allow the oligomer to undergo three-dimensional homo- and copolymerization with various monomers and oligomers. The hardening process follows either a free radical or an ionic polymerization or a polyaddition mechanism. OEM are harmless, miscible and readily react with a large number of monomers, oligomers and high polymers. The fact that acrylic oligomers have a regular structure allows to solve many problems arising in the synthesis of materials with required properties.

"Condensation telomerization" (the term was coined by A. Berlin[1, 2, 14]) is utilized in the synthesis of OEM and its analogs. This reaction is based on regular termination of chains during polycondensation by the introduction of compounds containing one functional group which reacts under given conditions and one (or several) functional groups which do not participate in the polycondensation reaction. This leads to the formation of oligomers with a regular structure, which enter into reactions of three-dimensional or "branched" polymerization through terminal or alternating groups. This is schematically illustrated by the synthesis of oligoester methacrylates:

$$R(OH)_n + R'(COX)_2 + CH_2=CY-COX \xrightarrow[80-120\,°C]{-HX}$$

$$\rightarrow (CH_2=CY-COO)_{n-1}ROCOR'CO \left[OR(OCOCY=CH_2)_{n-2}OCOR'CO \right]_m OR(OCOCY=CH_2)_{n-1}$$

X = OH, OCH$_3$, Halogen; Y = H, CH$_3$; R, R' = hydrocarbon residues of di- or polyalcohol and dicarboxylic acid.

This scheme reveals that, by varying the telogen-to-reagent ratio, the $\overline{M}_n$ value of oligoester methacrylates with terminal reactive groups may be changed in a wide range. A change in the nature of the hydroxy-containing component and the substituent at the double bond permits a variation in the number and nature of polymerizable groups. Condensation telomerization of aromatic and fatty dichloroformates with glycols or diamines, or the reaction of acrylic or saturated dibasic dichloroanhydrides with diamines yields numerous polymerizable analogs of OEM, containing carbonate, urethane and amide groups in oligomeric blocks[2, 13-15].

OEM and its analogs, like fumarate-maleates, may be hardened by irradiation or upon the addition of initiators and activators at 20 °C or at higher temperatures with the formation of strong, chemically inert and heat resistant (up to 200–250 °C) plastics. These plastics are widely applied as binding reinforced plastics, coatings, molding plastics, electrical insulators, sealants, adhesives, impregnating compositions, modifiers of rigid and elastic polymers, etc. Especially should be noted the capacity of OEM to incorporate large amounts of fillers and the use of OEM as temporal plasticizers and cross-linkers of commercial high polymers.

At present, Soviet industry produces five grades of low-viscous OEM containing, on the average, two terminal methacrylic groups (TGM-3, MGF-1, MGF-9, IDF-1, MDF-2) and three grades of high-viscous OEM with 4, 6 and 8 acrylic groups (TGF-11, 7-1, 7-20). Oligoesters composed of OFM and OEM and combined solutions of low-viscous (TGM-3, MGF-1) and high-viscous (di- and polyfunctional) OEM are of considerable interest. The former type of compositions (OFM-OEM) are becoming increasingly popular as monomer-free harmless binders, coatings and sealings, the latter type allows to regulate the viscosity and the properties of the systems forming three-dimensional polymers and copolymers[14, 15].

This brief description of the properties of unsaturated oligoesters indicates that they are promising starting materials for the production of free-foamed gas-filled polymeric materials. The very interesting possibilities of using composite foams based on OEM and commercial thermoplastics or elastomers should also be mentioned.

Many studies have been published on plastic foams based on OFM-styrene. The following synthesis is described in[16]. Unsaturated oligoesters on the basis of fumaric and maleic acids, phthalic anhydride and glycols, in the presence of small amounts of a polymerization inhibitor (hydroquinone), are mixed with styrene in ratios of 2.1 : 1 to 2.5 : 1; water, benzoyl peroxide and zinc stearate are then added. Sodium bicarbonate, a solution of maleic anhydride (or dicarboxylic acids or their anhydrides) and a small amount of an initiator (decomposing into radicals) are gradually added to the obtained mixture. The mixture of OFM, styrene and these reagents is stirred for 15 min and then poured into a mold where it is cured at 75 °C. In the reaction of carboxylic acids with the bicarbonate carbon dioxide, which foams the composition, is liberated, the initiating system (benzoyl peroxide – N,N-dimethyaniline) ensuring three-dimensional copolymerization of OFM with styrene. A plastic foam with a regular cellular structure (cell size 3–7 μm) and an apparent density of 55 kg/m^3 is thus obtained. It is well known that oxygen is one of the most effective inhibitors of free radical polymerization of unsaturated compounds. Therefore, the use of air as a blowing agent for unsaturated oligoesters, especially for systems with a short gel formation period is of considerable practical interest, provided that oxygen may be removed at the hardening stage. Such a procedure is described in patent[16] where it is proposed to stir mechanically with air the viscous mixture of OFM, styrene, initiator and activator

and then to pour the composition into a mold and harden the plastic foam under vacuum at elevated temperatures. Gas-filled materials with a partially opened cellular structure and an apparent density of $\gamma = 32-64$ kg/m^3 may be obtained by this technique.

Recently, oligoester-fumarate-maleate foams ($\gamma = 64$ kg/m^3) have been obtained on the basis
of porogenes[17]. To 100 parts by weight of a solution of OFM (based on adipic acid, dipropylene
glycol and maleic anhydride) in styrene or in vinyltoluene are added (parts by weight): 0.1–2.5
peroxide initiator (methyl ethyl ketone peroxide, H$_2$O$_2$); 0.001–0.2 Co^{2+}-containing activator
(e.g. Co-soap soluble in oil); 0.1–15 porogene [hydroxy-bis(benzenesulfonyl) hydrazide or tosyl
hydrazide], and 0.01–10 amino compounds (e.g. cyclohexylethylamine, dibutylamine, triethyl-
amine, triethanolamine, dimethylethanolamine, morpholine, methyl-substituted morpholines,
pyperidine, etc.) which activate the decomposition of porogene before gelation. Stirring of the
mixture before foaming allows to increase the viability due to additional aeration; foaming and
hardening is carried out at 18–32 °C. It should be noted that azo and hydrazo-compounds are
not only effective polymerization initiators but also porogenes of "OFM-monomeric" systems.
This considerably simplifies the formulation and processing of oligoester foams.

In order to prepare wood substitutes and materials with open-cell structures,
the so-called "water-filled" oligomeric foams are used[18]. These foams are obtained
by mixing the OFM-monomeric composition with water until an emulsion is formed
which is hardened after the addition of an initiator and activator. As a result, a white
rigid material is obtained which is a spatial network copolymer with uniformly
distributed micro-inclusions of water (2–5 μm). Optimal strength is reached at
50–60% of water, although the amount of water may be as high as 90%. At optimal
water concentration, the cell walls withstand cryolitic destruction till 34 K. At
7 mass % of water the apparent density of the material reaches 250–900 kg/m^3 and
it resembles natural wood in appearance and in some properties.

Freely foamed oligoester plastics are being increasingly used as light fillers in
load-carrying structures, as heat, sound and electrical insulating and buoyant materials.
This type of gas-filled material finds application in building, furniture and other in-
dustries.

However, the chemistry and technology of oligoester foams still does not receive
sufficient attention; information about their physicochemical properties, methods of
modification and technico-economical characteristics is incomplete. Especially, it
should be noted that the preparation of elastic and rigid foams on the basis of oligo-
estermethacrylates and "monomer-free" compositions of fumarate-maleate oligomers
has not yet been reported.

4.2 Polyfuran Foams

One of the most promising methods of foam preparation starts from polyreactive
oligomers which are obtained by polycondensation of 2-furfuryl alcohol, which is
subsequently hardened[19–23].

It is known that soluble reactive oligomers may be obtained by polycondensation of 2-furfuryl alcohol in the presence of acidic catalysts:

The products of furfuryl alcohol polycondensation contain ether bonds belonging to difurfurylic ether and possibly also to terminal groups of the oligomer[19].

The forming viscous oligomer is hardened at 120–140 °C in the presence of organic acids (e.g. maleic acid), or at room temperature using strong acids (sulphonaphthenic acids, hydrochloric and sulfuric acids, etc.) with formation of rigid heat-resistant reticular polymers. The mechanism of the hardening process has practically not been studied. Nevertheless, it may be assumed that in the presence of acidic catalysts the process of hardening involves cationic polymerization accompanied by opening of end furan cycles and interaction of the growing cation with multiple bonds of the oligomer.

Strong polyfuran foams ($\gamma = 20$ kg/m^3) of high heat and fire resistance, with closed-cell structure, are prepared from compositions containing 100 parts by weight of a mixture of furan oligomers, acidic catalyst (e.g. H_3PO_4, or its anhydride with H_3PO_4), blowing agent (NaHCO$_3$, n-hexane, chlorofluoroalkanes, etc.) and less than 100 parts by weight of active filler (lignine derivatives, ash, talc, wood meal and others)[19]. Composition-mixtures containing furfuryl alcohol, 10–75% of furfuryl alcohol prepolymer and 10–80% of the oligomeric product of urea cocondensation (thiourea or melanine with formaldehyde and furfuryl alcohol) are used as furan oligomer mixtures. Foaming and hardening is performed without external heating and takes several minutes.

Strength, fire resistance and apparent density of polyfuran foams depend on the composition of the mixture, amount and type of filler. By varying these parameters, foamed materials may be obtained with $\gamma = 20-100$ kg/m^3 or higher. The fire resistance of these materials is better than that of foamed phenoplastics and polyisocyanurate foams, and the mechanical strength is higher than that of carbamide foams.

4.3 Other Types of Oligomeric Foams

Promising raw materials for the preparation of oligomeric foams are methylol derivatives of aliphatic ketones, e.g. of acetone. In the late 1930's it was shown that methylol derivatives of acetone and methyl ethyl ketone may be converted into transparent, strong, non-melting, insoluble polymers in the presence of basic catalysts and peroxides at 20–80 °C. The mechanism of this reaction is still unclear. Depending on the conditions, the process presumably follows either a radical-chain pathway ("tautomeric polymerization"), or a step-wise polycondensation, accompanied (or not) by the polymerization of dehydration products of α, β-unsaturated ketones. Irrespective of the complexity of the conversion of ketone methylol derivatives in basic media, this conversion invariably results in the formation of network copolymers and proceeds through stages of flexible polyreactive oligomers.

According to the foregoing discussion it may be assumed that in the polycondensation of mono-, di- and trimethylol ketone mixtures, the conditions for low-temperature hardening and foaming of the appropriate precondensates may be found. Indeed, in recent years many patents have claimed for methods of preparation of plastic foams from methylol derivatives of aliphatic ketones. The techniques involved are ususally as follows: Mono-, di-, trimethylol ketones or their mixtures are subjected to polycondensation at mild temperatures (70 °C), then a blowing agent (up to 7–10% of fluorocarbons, low-boiling hydrocarbons or alcohols) and a catalyst (40–60% aqueous alkaline solutions) are introduced and, if necessary, silicone surfactants, plasticizers and powder fillers are added. Mixtures are poured into molds where they are foamed and hardened at 70 °C[24, 25].

Among promising oligomeric foams one should also mention 2-pyranyl foams for a discussion of the chemistry, technology and properties of these materials see[26]. Finally, several works concerned with the preparation of foams from cumaron-indene[27] and aniline-formaldehyde oligomers have been reported[28].

4.4 Multicomponent Oligomeric Foams

One of promising methods of modifying polymer properties is mixing polymers with different chemical structure and consequently with different properties. Such "mixing" or "combining" makes it possible to prepare polymeric materials which combine properties of all the components of the mixture.

Studies of preparation and properties of the so-called Simultaneous Interpenetrating Networks (SIN) occupy a special position among these works. SIN is a complex system of two or more three-dimensional network polymers which are chemically not bonded but are inseparable due to mechanical entanglement of chains. A detailed description of the preparation and properties of SIN is given by Lipatov and Sergeeva[29].

In this review foams on the basis of SIN systems and polymer-oligomeric systems are referred to as multicomponent oligomeric foams.

Of recent works we would like to mention first of all the preparation of multi-component phenolic-based urethane foams[30−34]. These materials combine the advantages of the initial components (fire resistance of phenolic and high strength of polyurethane foams), whereas their shortcomings (friability and high water-absorption of phenolic and flammability of polyurethane foams) are suppressed to a considerable extent.

The most detailed study of the physicochemical formation of phenolic-based urethane foams has been carried out in the USA by Schafer and co-workers (Ashland Chem. Corp. Columbus, Ohio) and K. C. Frisch and co-workers (Polymer Institute, Detroit, Michigan)[34].

Five types of phenolic polyols were examined in the latter work, the properties of two of these polyols are given in Table 1.

The method of preparation involves mixing of polymeric isocyanate ("crude" MDI) with the polyol, surfactant, catalyst and blowing agent, trichlorofluoromethane (fluorocarbon F-11B). The amounts of the latter and of the catalyst were chosen so that the apparent density was

Table 1. Physical properties of phenolic polyols[34]

Property	Type of phenolic polyol	
	III	IV
Hydroxy number (mg KOH/g)	567	515
Water content (%)	0.467	0.658
pH value	4.25	5.20
Specific gravity at 20 °C	1.2369	1.1808
Viscosity at 25 °C (cps)	49.800	10.720
Refractive index (n_d^{25})	1.5910	1.5700

32 ± 3 kg/m^3 and rise time was from 75 to 180 s. The foams were then hardened for 24 h and stored at room temperature for at least a week. Formulations and properties of the obtained foams are presented in Table 2.

Table 2. Formulations[a] and properties of phenol-urethane foams based on various types of phenolic polyols[34)]

Component content, composition and foam properties	Type of phenolic polyol			
	I	II	III[b]	IV[b]
	Composition			
Polyol(s) (g)	100	100	100	100
DC-193 (g)	3.0	3.0	3.0	3.0
Freon 11-B (g)	31	35	32	32
DABCO R-8020 (g)	0.15	0.15	0.8	0.8
	Composition properties			
Cream time (s)	44	50	30	44
Rise time (s)	90	146	88	81
Tack free time (s)	93	148	74	74
	Foam properties			
Apparent density (kg/m^3)	34.6	30.4	30.4	30.5
Friability (%)	14.0	9.8	3.4	8.0
Oxygen index	24.75	24.75	24.88	24.00
Compressive strength (kg/cm^2):				
parallel to rise	4.83	2.59	2.00	1.96
perpendicular to rise	2.28	1.12	1.22	0.5

[a] Isocyanate index for all formulations was 105
[b] Samples were hardened for 3 days before testing

It has been found that with increasing amount of surfactant the friability decreases, and at optimal amount of surfactant (2.5–3%) the friability of foams does not exceed 10%; by changing the type of catalyst, the friability is reduced from 16 to 7.0%. The oxygen index of phenolic-based urethane foams increases noticeably with the introduction of flame-retardant additives. Thus, the addition of 10% of Firemaster T23P [tris-(2,3-dibromopropyl) phosphate] to the polyol-III formulation increases the oxygen index from 24.88 to 27.60 and the addition of 10% of Furol-76 (vinyl phosphate oligomer) to the oligomer-V formulation increases the index from 24.00 to 27.06. These materials reveal marked humid aging characteristics. After aging for 14 days at 70 °C and 100% relative humidity the foam based on phenolic oligomer-V increases its volume only by 1.9%, whereas the same parameter for polyurethane foams is 5.1–7.3%.

In the USSR two groups of scientists[35)] and [36, 37)] obtained semi-rigid phenolic-based urethane foams by foaming compositions containing polyester and resol phenolic oligomers. These foams are prepared by similar formulations from a mixture of polyester, diisocyanate, 2,2,2-trichloroethyl phosphate, phenolic oligomer and catalyst. Carbon dioxide formed in the reaction of a diisocyanate with water (present in the phenolic oligomer) acts as the blowing agent. The main characteristics of these multicomponent oligomeric foams of grades SFUP[36, 37)] and SEP-1[35)] are given in Table 3 together with the properties of their "prototypes" phenolic (FRP-1) and urethane (PPU-3S) foams. SFUP and SEP-1 foams are used for the manufacture of building sandwich constructions.

Table 3. Properties of "combined" foams and of their prototypes

Property	Foam grade			
	SFUP[a]	SEP-1[a]	FRP-1	PPU-3S
Time (s)				
starting	25	20–30	60–90	20–25
rising	200–90	215	260	160
gelation	110	150	210	90
curing	350	380	300	400
Apparent density kg/m^3	40–120	60	60	60
Strength (kgf/cm^2):				
compressive	0.8–10.1	1.7	0.8	3.0
tensile	2.7–7.6	4.0	1.6	6.5
Modulus of elasticity (kgf/cm^2):				
in compression	30–130	–	80	120
in tension	60–192	195	100	210
Technological shrinkage (%)	0.3	0.5	1.5	0.2
Water absorption after 24 h (vol.%)	2.5–3.5	3–5	10	1–2

[a]   Ratio of phenolic to polyester oligomer 1:1

For IR spectra of all components of multicomponent oligomeric compositions and of SFUP, FRP-1 and PPU-3S foams see[37]. It has been established that the IR spectra of SFUP do not exhibit any new absorption bands in relation to the spectra of PPU-3S and FRP-1 foams and the spectrum of a mechanical mixture of these foams. These data indicate that no chemical interactions occur between these combined three-dimensional systems. This is in good agreement with modern notions about the mechanism of SIN formation[29].

A discussion of the formation chemistry and the molecular mechanisms responsible for the properties of multicomponent oligomeric foams is not the subject of this survey. We only emphasize here the complexity of these problems which is connected with the morphology of gasfilled polymers (see the six levels of structural organization, Chap. 3.2).

Several other types of multicomponent oligomeric foams have been proposed: Foams based on mixtures of resol phenolic oligomers with novolac oligomers[4, 38–41], with urea oligomers[4, 42–44], with polycarboxamides[45], with polyacrylonitrile[46, 47], with polystyrene[48], with bitumens[49]; multicomponent oligomeric foams based on polyurethanes and urea oligomers[50–53], copolymers of styrene and acrylonitrile[54], monomers and oligomers of acrylate and methacrylate[55]; foams composed of mixtures of urea oligomers with melamine-formaldehyde oligomers[56–58], with urethanes[44], with poly(vinyl chloride)[59]. And finally foamed plastics based on mixtures of furan and urethane, phenolic and urea oligomers[21, 22, 60, 61].

Foamed polymers based on mixtures of resol phenolic and carbamide oligomers; resol and novolac oligomers; epoxy and novolac phenolic oligomers have been produced industrially in the USSR for several years[4, 38–40, 62].

In our opinion, multicomponent oligomeric foams provide a variety of chemical-technological and technical-economical possibilities which let expect practical utilization in the near future.

To conclude this chapter, it should be emphasized that all studies associated with the development of new types of foams and with the modification of existing materials should be carried out taking into account the ecological problems on our planet. The first serious attempts towards such an approach with respect to the production and use of oligomeric foams have already been made by Baumann and Schuur[57, 63, 64].

5 New Data on the Morphology of Oligomeric Foams

5.1 The Quantitative Concept of Uniformity

As already mentioned, only isotropic RO foams are examined in this paper. However, the concept "isotropy" may be applied to any heterogeneous material, including polymeric foams, only with stringent reservations. It is well known that, along with integral foams as typical representatives of anisotropic foams, isotropic ("classical") foams always display anisotropy of structure and consequently of properties. The specific gravity, as well as cell structure and shape of RO foams prepared by free foaming always vary to a certain extent along with the width and height of a foamed article. The number of open and closed cells sharply differs at the center and at the periphery of a foam panel.

Knowing all this, we nevertheless often use the terms "uniform foam" and "uniform foam-structure" without providing them with a definite scientific meaning but rather relying on our visual or even on esthetic notions.

The concept "uniformity" of a material, however, has the following quantitative definition: the coefficient of uniformity of a material, from the point of view of probability, is the ratio of some property (technical characteristic) of the material at a given value of probability (for a known integral distribution curve) to the arithmetic mean value of this property.

For a normal distribution, the coefficient of uniformity K_1 is given by

$$K_1 = \frac{M - 3\sigma}{M},$$ (1)

where M = arithmetic mean value of a property in a given batch of samples and σ = the mean-square root deviation.

Evidently, the more "uniform" the material, the closer K is to unity.

The distribution of some important technical characteristics of plastic foams, e.g. strength, is closer to the Weibull than to the Gaussian distribution due to "non-uniformity". Kozlov[65] obtained the analytical expression for the coefficient of uniformity K_2 for the Weibull distribution and used it to estimate the uniformity of phenolic foams.

The integral function of the Weibull distribution may be written as

$$F(x) = 1 - e^{-(x/x_1)^m}$$ (2)

where x = the measured quantity (e.g. the strength of the foam in a series of measurements); x_1 and m = parameters.

By setting a certain value α on probability $F(x)$ we have

$$d = 1 - e^{-(x_2/x_1)^m} \tag{3}$$

and then taking logarithms we obtain

$$x_2/x_1 = [-\ln(1-\alpha)]^{1/m} \tag{4}$$

Introduction into the left and right-hand side of Eq. (4) of value $1/b_m$ so that $a = x_1 b_m$, where a is the mean value, gives

$$\frac{x_2}{x_1 b_m} = \frac{1}{b_m} \sqrt[m]{\ln \frac{1}{1-\alpha}}$$

or   (5)

$$\frac{x_2}{a} = \frac{1}{b_m} \sqrt[m]{\ln \frac{1}{1-\alpha}}$$

If the uniformity of the plastic foam relative to strength characteristics is determined, the left-hand side of Eq. (5) is the ratio of strength at given probability α to the mean value of strength and

$$K_2 = \frac{1}{b_m} \sqrt[m]{\ln \frac{1}{1-\alpha}} = \frac{1}{\Gamma\left(1 + \frac{1}{m}\right)} \sqrt[m]{\ln \frac{1}{1-\alpha}} \tag{6}$$

where

$$\Gamma\left(1 + \frac{1}{m}\right)$$

is a gamma function.

Hence, Eq. (6) determines the coefficient of uniformity through the parameters of the Weibull distribution.

Kozlov[65] compared Eqs. (1) and (6) for tensile tests of 56 specimens of phenolic foams based on resol oligomers (FRP-1 grade). In order to exclude the effect of differences in the apparent density of samples which were cut from the same panel, specific strength characteristics were used, i.e. the ratio of strength to specific gravity.

The experimental data for calculations according to Eq. (1) are

$$M = 0.0376 \frac{kg/cm^2}{kg/m^3} \; ; \; \sigma = 0.0081 \frac{kg/cm^2}{kg/m^3} \; ; \; K_1 = 0.338$$

Figure 1a shows the experimental function of reliability for a given batch of specimens. This graph on "probability paper" corresponding to a normal distribution demonstrates that a straight line cannot be drawn through these points, i.e. the observed distribution is not "normal". When plotting this graph on "probability paper" corresponding to the Weibull distribution (Fig. 1b), the points fit to a straight line. For data of Fig. 1b we have

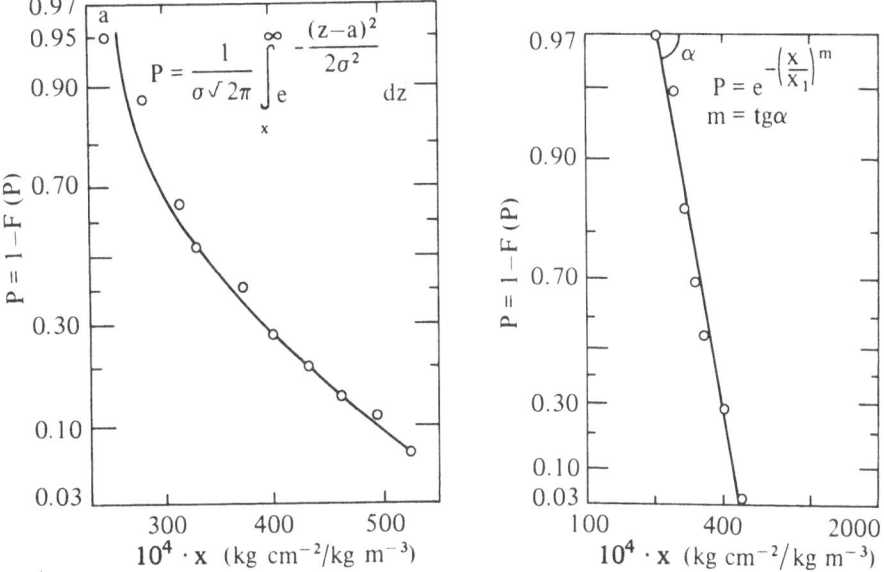

Fig. 1 a, b. Plots for the determination of uniformity coefficients (reliability function) of phenolic foam samples during tensile tests (a) normal (Gaussian) distribution; (b) Weibull's distribution[65]

$$x_1 = 0.0390 \frac{kg/cm^2}{kg/m^3} \; ; \; \alpha = 80°; \; tg\,\alpha = 5.67$$

$$m = 5.67; \; \Gamma\left(1 + \frac{1}{5.67}\right) = 0.925$$

Let us now determine the numerical value of K_2 from Eq. (6). Since the numerator in Eq. (1) equal to $(M - 3\,\sigma)$ leads to intervals of grouping of strength, for which the probability of appearance is equal to 0.003, we shall take the values of α also equal to 0.003:

$$K_2 = \frac{1}{0.925} \sqrt[5.67]{\frac{1}{1 - 0.003}} = 0.388$$

Hence, the value of K_2 exceeds K_1 by 15%. Therefore, the observed distribution of the specific strength of the phenolic foam corresponds to the Weibull distribution and the given batch of specimens is uniform in terms of probability according to Weibull and non-uniform according to Gauss.

It should be noted that up to now, quantitative estimates of the uniformity of gas-filled plastics have not been carried out, although the importance of this concept for the estimation of the morphology of foamed plastics is obvious. Indeed, for researchers and technologists the quantitative determination of the uniformity of foams should become one of the basic criteria of the quality of articles, along with other morphological parameters. On the other hand, for theoretical calculations, it should be remembered that the coefficient of uniformity is a "brutto characteristic" of foam heterogeneity and incorporates the total "non-uniformity" of the distribution in bulk of the material of all other morphological parameters: apparent density, quantity, size and shape of cells, number of open (or closed) cells, etc.

In further discussions we will frequently use another morphological parameter, the volume fraction. It will be recalled that, in the general case, a plastic foam is not a two-phase but a three-phase system "solid — gas — liquid".

Using the concept "volume fraction" we may write

$$\vartheta_p + \vartheta_w + \vartheta_g = 1 \tag{7}$$

where ϑ_p, ϑ_w and ϑ_g are the volume fractions of the polymeric phase, water and gas in the foam and

$$\vartheta_p = \frac{m}{V \cdot \gamma_p} = \frac{\gamma}{\gamma_p} \; ; \; \vartheta_w = \frac{m_w - m}{V \cdot \gamma_w} \; ; \; \vartheta_g = \frac{m_g}{V \cdot \gamma_g} \tag{8}$$

where V is the geometrical volume of the foam; γ_p, γ_w, γ_g and γ are the specific gravities of the polymeric phase, water and gas and the apparent density (volume weight) of the foam respectively; m, m_w and m_g are masses of dry foam, wet foam and of the gas in the foam cells.

5.2 Microcellular Structure

Depending on the type of initial oligomer, composition and technology of foaming, the upper limit of the cell size is in the range of several millimeters. As for the minimum size of cells in oligomeric foams, it was considered not so long ago to be not smaller than several dozens of microns. However, systems of gas cavities in RO compositions were described several years ago, the minimum size of which were only fractions of a micron (μm), i.e. by 2–3 orders of magnitude smaller than previously observed. Such structures with dimensions of several microns or less were designated by the author as microcells.

Micromorphological systems in the form of isolated closed and open microcells of strictly spherical shape and also agglomerates of microcells in the form of "coils" and arranged in one line (parallel to rise) have been detected; through micro-holes and micro-windows, covered with a film in the macrocell, walls (membranes) were discovered by scanning electron microscopy (Spector and Shutov[66]) in resol phenolic oligomeric foams. For a detailed description of microcells, their statistical analysis and micrographs see[4].

The existence of microcells in the structure of oligomeric foams was confirmed by studies on other types of oligomeric foams: phenolic[8, 9, 67], polyurethane[4, 68–70] and urethane-phenolic multicomponent oligomeric foams[36, 37].

In the following the microstructure of polyurethane foams based on polyesters (grade PPU-307)[12, 71] will be described.

Figure 2 shows the results of a mercury-penetration study of two PPU-307 specimens clearly differing in apparent density, $\gamma = 40$ kg/m^3 (Fig. 2a) and $\gamma = 500$ kg/m^3 (Fig. 2b). The analytical shape of the differential distribution curves of the volume fraction of all cells $\vartheta = \vartheta_g$ for both specimens is given by

$$\frac{\Delta \vartheta}{\Delta \ln R} = C_1 \frac{\vartheta_p}{\vartheta_p^{max}} \exp\left(-\frac{\vartheta_p^2}{\vartheta_p^{max^2}}\right) + C_2 \frac{1}{\ln \sigma \sqrt{2\pi}} \exp\left(-\frac{\ln R - \ln R^{max}}{2 \ln^2 \sigma}\right) \tag{9}$$

where ϑ_p = volume fraction of the polymeric phase in the foam, ϑ_p^{max} = 0.368 = volume fraction of the polymeric phase corresponding to the closest sphere packing, R = the cell radius, R^{max} = abscissa of maximum R, σ = standard deviation of ln R, C_1 and C_2 = empirical positive constants.

The first term in Eq. (9) describes the contribution of microcells ($R \leqslant R^{max}$), the second one that of macrocells ($R < R^{max}$).

Finally, the cell size distribution is not Gaussian but a normal logarithmic one. This implies that, according to the theory of statistical analysis, the totality of cells consists of several, at least two, groups of cells. The cells of these groups markedly differ both in size and in number. This is confirmed by a further analysis of the data available (Fig. 2).

The specific surface of the specimens was calculated according to

$$S = \frac{6 \sum\limits_i \nu_i \lambda_i^2}{D\gamma \sum \nu_i \lambda_i^3} , \tag{10}$$

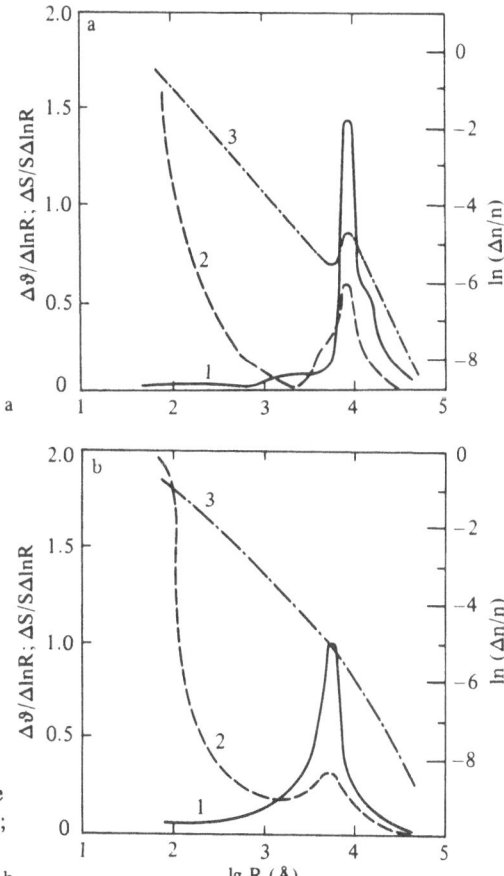

Fig. 2 a, b. Differential curves of the distribution of cell radii for rigid PUR foam based on polyester (grade PPU-307) of the apparent densities 40 kg m^{-3}(a) and 500 kg m^{-3}(b). 1: distribution of the volume fraction of cells $\Delta \vartheta/\Delta \ln$ R; 2: distribution of the relative specific surface area $\Delta S/S \Delta \ln$ R; 3: distribution of the relative number of cells ln $\Delta n/n$

where D = most probable size of cells, γ = apparent density of the foam, ν_i = frequency of appearance of cells with size D_i, $\lambda_i = D_i/D$.

The ratio of the sums in the numerator and denominator of Eq. (10) is equal to unity only for the closest sphere packing of monodisperse cells; for all other types of packing and for polydisperse cells this ratio is always less than unity. Differential curves of the specific surface distribution over cell sizes (Curves 2 in Fig. 2a, b) are described by Eq. (11):

$$\frac{\Delta S}{S \Delta \ln R} = C_1 \exp\left(-\frac{\pi^3 R}{2\, R^{max}}\right) + C_2 \left(\frac{R}{R^{max}}\right)^2 \exp\left[-\left(\frac{R}{R^{max}}\right)^2\right] \tag{11}$$

The first term represents the contribution to the specific surface of microcells (R < 1000 A), the second one that of macrocells (R > 1000 A).

As can be seen from the differential curves of the cell volume fraction distribution (Curves 1 in Fig. 2a, b), the microcells occupy a relatively small volume of the foam: 5% for γ = 40 kg/m^3 and 11% for γ = 500 kg/m^3. They are, however, the most numerous group of cells (Curves 3, Fig. 2). Cells with a radius of 100 A or less make up about 50% of all cells; the entire population of microcells (the size of which is equal or less than 1000 A) is by at least three orders of magnitude higher than that of macrocells. Due to the very small size of microcells and their large number, the specific surface S of foams for the light specimen (γ = 40 kg/m^3) is as high as 100 m^2/g and for the heavy specimen (γ = 500 kg/m^3) S = 30 m^2/g.

It follows from Fig. 2 and Eq. (9) (second term) that macrocells occupy most of the foam volume (> 70%). It is of interest that an increase of the apparent density by an order of magnitude from 40 to 500 kg/m^3 reduces the most probable radius of macrocells only by a factor of 2. The contribution of macrocells to the total specific surface value is 2–3 m^2/g (for γ = 40 kg/m^3) and 0.5 – 1 m^2/g (for γ = 500 kg/m^3).

Using Aleksandrov's formula[72], the average wall thickness $\bar{\delta}$ of macrocells is estimated as

$$\bar{\delta} = \bar{D}\left(\frac{1}{\sqrt[3]{1 - \gamma/\gamma_p}} - 1\right), \tag{12}$$

where $\bar{D}$ = average cell diameter, γ_p and γ = specific gravity of the polymer and apparent density of the foam respectively.

It has been found that the value $\bar{\delta}$ for γ = 40 kg/m^3 is by two orders of magnitude smaller than the macrocell diameter ($\bar{\delta}$ = 0.03 μm); for the heavy foam $\bar{\delta}$ = 0.2 μm.

According to these results the following important statement can be made: the structure of oligomeric foams is a matrix system of thin polymeric films with two groups of cells, sharply differing in size and number and enclosed into one another.

This statement is additionally substantiated by data on dielectric properties of foams which will be discussed in Chap. 7.1.

Findings of the author and his colleagues[4, 8–10, 12, 37, 66–71] concerning the existence in the structure of oligomeric foams of microcell systems were later confirmed by Soviet scientists[11] (see Chap. 6.3), as well as by a group of British researchers, Lowe, Barnatt, Chanley and Dyke[73] in a study of rigid closed cell phenolic foams expanded by fluorocarbons and pentane.

In all photographs obtained by a scanning electron microscope Lowe and coworkers have observed interstices between macrocells, the cross section of which is 1.5–2 μm. A large number of interstices of approximately the same size have also been observed in the structure of phenolic foams[4, 8, 9, 67]. Lowe et al. deduced the presence of these interstices from the well-known drift effect observed during measurement of the fraction of closed cells by the ASTM D2856-70 method. Thus, in a series of measurements of three phenolic foam specimens the first measurements yield 94, 81 and 30% of closed cells; after 15 min the values decrease to 73, 61 and 24%, respectively. The drift effect is interpreted as a process of gradual penetration of gas into interstices between cells and is not connected with the mechanical destruction of cells during measurements since the effect is reproducible for the same spicemens[73]. As for the shape of interstices, the authors assume that the development of "fine cracks arises from a lack of bonding between individual cells"[73].

On all photographs micromorphological formations about 1 μm in diameter are clearly seen; Lowe et al. call them "microholes" (in our terminology, microcells). At high enlargement (5500 X) the photograph exhibits examples of both explosions and implosions on the same cell together with cyst-like protuberances which have not exploded. Our observations of microcell morphology in phenolic foams are very similar[4]. In open-cell foams Lowe et al. observed macroholes or holes together with microholes. In a private communication (cited by Rossmy et al.[74]) Lowe has reported 1 μm microholes in the structure of rigid PUR foams.

Let us now consider the differences between the absolute sizes of microcells reported by Lowe et al. and those found in our experiments. In our first publications[8, 9, 66] and in a monograph[4] the minimum size of microcells in phenolic foams has been stated to be 6 μm. The value was obtained for large-cell heavy (γ = 200 kg/m^3) foams. In later works[5, 12, 69, 70] light phenolic foams (γ = 40 kg/m^3) were studied which are comparable in their specific density with the materials investigated by Lowe et al. (γ = 35 kg/m^3) and microcells were found to be much smaller (1.0–1.5 μm). Note, that this value corresponds to the most probable cell size (abscissa maximum on the differential distribution curve (Fig. 2)). According to our data[12, 69] the structure of phenolic foams includes not less than 20% of microcells with sizes smaller than 1.0 μm.

Lowe et al.[73] did not carry out a statistical analysis of the relative number of microcells and their size distribution in the bulk of the material. It should be mentioned, however, that they do not consider microcells as the most frequently occurring group of cells, i.e. as a morphological system, in contrast to our views. For a discussion of the difference in quantitative data see Chap. 5.3).

Several other experimental findings support the existence of a microcellular structure in oligomeric foams. Thus, Oween and Denis[75] observed an "anomalous" pattern (in the expression of the authors): for certain types of silicone surfactants the liquid foam system consists of gas bubbles the sizes of which differ by several orders of magnitude. The possibility of formation of very small gas bubbles after a marked reduction of the surface tension coefficient in polyurethane formulations has been reported by Dubyaga and Tarakanov[76].

Finally, a number of theoretical works (see e.g.[77] and[78]) predict the existence of gaseous cells differing in size by several orders of magnitude in non-Newtonian liquids[78] and low-viscous media[77].

Hence, all the above-mentioned data fit into a quite consistent system of new notions about the morphology of oligomeric foams.

5.3 The Mechanism of Microcell Formation

Let us now examine the formation of such microstructures. Lowe and co-workers[73] consider that the formation of microcells is caused by a rupture of macrocell walls by formaldehyde vapors. Indeed, oligomer polycondensation of 2-methylolphenol in the presence of acidic catalyst may partially proceed via 2-hydroxydibenzyl ethers:

In addition to water, formaldehyde the boiling point of which is only about
$-20\,°C$ (anhydrous form) is formed in this reaction. The possibility of water vapor
formation has been rejected as being the cause of the production of microcells since
the temperature evolved in foaming does not exceed $85\,°C$.

Naturally, the water vapor may cause the formation of microcells in other
types of phenolic compositions, although the formation of microcells due to formaldehyde vapor is undoubtedly quite possible for any phenolic foam. Lowe et al.
used low-boiling liquids (fluorocarbons F-11 and F-113 and pentane) as blowing
agents whereas we used a different blowing system: hydrogen formed in the interaction of aluminum powder with orthophosphoric acid in the presence of benzene-
sulfonic acid. During foaming of our composition the temperature increases to
$105-110\,°C$, which is quite sufficient to evaporate both polycondensational water
and the considerable amount of water present in the initial resol oligomer. Therefore,
microcell formation due to water evaporation is presumed to occur in our system.

Differences in the gas-forming systems and in the temperature may also explain
the variations in the relative number of microcells in the foams.

Convincing evidence of the effect of water vapor on the morphology of oligomeric foams
was provided by Rossmy and co-workers[74]. Although the authors studied flexible PUR foams
(on the basis of polyesters) rather than rigid phenolic foams and their aim was to elucidate the
mechanism of cell opening, in our opinion the results obtained in this study are of a more general
nature and substantially extend our knowledge of the formation of cellular structure (including
microstructure) in foamed plastics based on polyreactive oligomers.

Rossmy and co-workers established that the temperature of the composition at which cell
opening occurs is practically independent of the concentration and type of the surfactant and
catalyst, isocyanate index and water content and is close to $100\,°C$. The authors suggested that
the process of cell opening is connected with water evaporation in the formulation, since of all
the components only water has a phase transition at $100\,°C$. In order to prove this assumption,
Rossmy and co-workers carried out a rather elegant experiment: the temperature of cell opening
was determined at various external pressures ($460-1060$ mm Hg) of the foaming composition.
The results of this experiment (Fig. 3) confirmed their suggestion: the experimental points practically coincide with the temperature dependence curve of water vapor pressure, especially if one
takes into account that the accuracy of the determination of the cell opening temperature was
$±2\,°C$.

Thus, one of the possible causes of cell opening has been established by a direct
physical experiment. However, the authors proposed two further possible mechanisms
of the phenomenon.

1. "Water in the foam mixture (which, in the early stages of foaming is homogeneously distributed as hydrate complexing ether- and OH-groups) separates in the form of small droplets
   when the increase of temperature leads to a splitting of the complex-forming hydrogen
   bonds. These small water droplets will be surrounded by a microatmosphere of water

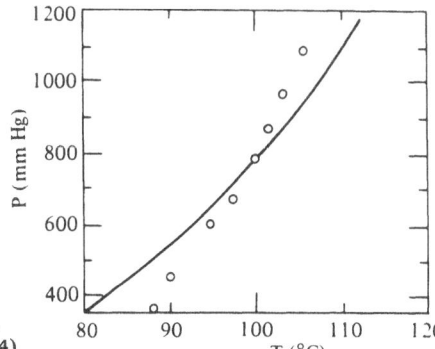

Fig. 3. Dependence of cell opening temperature (*points*) and water vapor pressure (*solid curve*) on external pressure during foaming of flexible PUR foam (water content: 4 parts by weight)[74]

vapor. The boiling of water leaves small holes in the membranes which permit an approximate equalization of pressure and reduce the expansion rate. Cell opening then follows by the sucking of the membrane into the plateau border. Of course, it can also be imagined that spontaneous boiling of water droplets might also disturb the films mechanically in a way leading to rupture and, subsequently, in flowing of the membranes."

2. "Water is an essential component of the colloidal structures stabilizing the foam. Separation of water followed by boiling reduces the strength of these structures to cause partial defoaming."

Experimental findings of Rossmy et al. substantiate the first mechanism. It has been established that only 18–24 vol% of water is vaporized during foaming. An IR spectroscopic study reveals that the chemical structures of walls and struts differ noticeably. Thus, the intensity ratio of absorption bands, 2970/2270 cm^{-1} (methyl groups of polyether to isocyanato groups), is 2.85 for cell membranes whereas for struts it is 3.27; the 1595/2270 cm^{-1} ratio (C=C bonds to isocyanato groups) is 2.49 and 3.17 for membranes and struts respectively. This difference in the isocyanate content causes a difference in temperature between membranes and struts and consequently the former lose more water during foaming. Thus evaporation of water leads to cell opening rather than to defoaming.

The authors of[74] commenting on Lowe's private communication (see above) about the existence of microholes in the structure of rigid PUR foams put forward the following assumption: "in a rigid system the probability for a flowing of the membranes due to suction of plateauborders is by far smaller because the viscosity of the film-forming material is higher. Of course, we cannot take the holes in the membranes of a rigid foam as a proof for a temporary existence of holes in flexible foams. And it is not certain that boiling water is the reason for the occurrence of holes in the rigid foam systems either, although both systems under investigations contained water."

Hence, one of the most probable reasons for the formation of open cells, including microcells, due to water vapor is given in[74]. The mechanism described there, however, does not explain the formation of closed microcells which we observed in the structure of rigid phenolic foams[4]. The formation of these microcells is apparently connected with specific features of foaming and hardening kinetics in oligomeric compositions.

In our foam composition "phenolic oligomer – hydrogen", liberation of gas due to its oversaturation in the systems occurs even at room temperature (Fig. 4). However, the foam begins to rise at about 50 °C (region I). During the following 60 s (region II) the foaming rate is constant due to the constant rate of temperature increase (up to point B). As a result, a dynamic equilibrium between sorption of gas (H_2) by the composition and gas desorption from the composition into bubbles is established in the AB interval (region II). According to the well-known mechanism of Saunders and Frisch[78], under conditions of equilibrated saturation concentration,

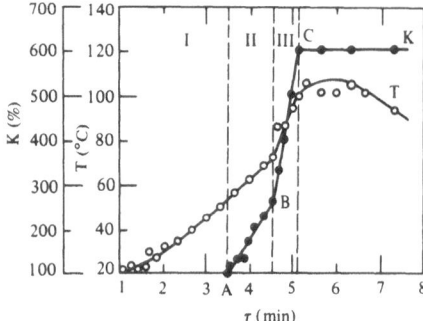

Fig. 4. Variation of foaming degree K and temperature T during foaming of phenolic foam FL-l ($\gamma = 200$ kg m^{-3})[4]

new bubbles are not formed and the foam rises only due to the expansion of the bubbles formed at point A. Evidently, it is these bubbles that form macrocells.

At point B (75 °C) the rate of foam rise sharply increases indicating a noticeable increase in the volume of the blowing gas at this point (region III). Only at dynamic gas equilibrium, however, the increase of the gas volume in the system is not related to the formation of new bubbles. In our case, the equilibrium is disrupted, as can be seen from a change of the temperature curve slope. Indeed, at point B the rate of the temperature increase markedly changes, most probably on account of the contribution of the exothermal reaction of oligomeric matrix hardening. When the dynamic gas equilibrium is distorted, the gas volume increases not only through diffusion of gas into existing bubbles and gas expansion within these bubbles but also through formation of new bubbles. A "second wave" of foaming starts at point B.

Hence, the formation of microcells in the structure of phenolic foams is due to the fact that, at a certain stage of foaming (region III), thermodynamic conditions are created for the liberation of gas from the oversaturated mixture (formation of a new portion of bubbles). These bubbles expand according to the same laws as bubbles formed at point A, but they do not attain the same sizes as the macrobubbles. This is ascribed to the fact that at point C the viscosity of the foam system has increased to such an extent that the gas pressure in all bubbles (both macro- and microbubbles) is insufficient to cause a further expansion and the foam does not rise further.

Berlin[6] already indicated more than 25 years ago, that the process of gas evolution in "phenolic oligomer – hydrogen" foaming systems occurs much more readily than in those where other gases or gaseous substances (nitrogen, carbon dioxide, fluorocarbons, etc.) are evolved, because:

1. during hardening of the phenolic oligomer when the rigidity of the three-dimensional network drastically increases, permeability, sorption and diffusion coefficients of the gases evolved decrease much more rapidly than in the case of thermoplastic linear polymers;

2. with increasing temperature the solubility of hydrogen in oligomers and polymers decreases more rapidly than the solubility of other gases and gaseous substances.

The proposed mechanism may be generally applied to all foam systems based on polyreactive oligomers. Thus, the formation of microcells in PUR foams and in phenolic-based urethane foams may also be explained by this mechanism. Of course, low-molecular compounds are also formed in these systems as a result of polycondensation, but they do not sensibly affect microcell formation. On the other hand, in the cell opening process, the role of these substances (including water) is of primary importance which is in agreement with Rossmy et al.[74].

Recent studies on the morphology of oligomeric foams have thus considerably extended our knowledge of the structure of these materials. For the last few decades, macrocells with dimensions of the order of $0.1-1$ mm were considered as the only morphological unit. At present it can be stated that in oligomeric foams, along with macrocells, another type of structural organization is present – microcells, which are by $2-3$ orders of magnitude smaller than macrocells.

To conclude this discussion of the microcellular structure of oligomeric foams we would like to emphasize that the existence of microcells should not be considered as "defectiveness" in the morphology of these foams caused by "distortions" under the prevailing technological conditions of foaming or by "mistakes" in the selection of components. On the contrary, the analysis of physicochemical regularities of foaming of oligomeric compositions proves that the appearance of microcells is connected with specific features of the chemical mechanism of cross-linking in polyreactive oligomers (formation of low-molecular compounds) and with the specificity of foaming kinetics (disruption of the dynamic gas equilibrium in the "oligomer-gas" system at the final stages of cellular structure formation).

5.4 Specific Surface and Degree of Dispersion

The specific surface of (phenolic) plastic foams was first measured by Schauer, Truxa and Spitzer in 1967 using the BET method[79]. They obtained a value of about $2 \, m^2/g$.

The high specific surface of PUR foams (Fig. 2) reaching $140 \, m^2/g$ ($\gamma = 40 \, kg/m^3$) was mentioned above. According to our data the value of S is $5-10 \, m^2/g$ for phenolic foams, whereas Fedodeev and Litvinova obtained much higher values, $425 \, m^2/g$, corresponding to a degree of dispersion of $8.4 \cdot 10^7 \, cm^{-1}$ [11] (these data will be discussed on p. 40 together with the problem of water absorption, see Table 6). These great differences are presumably due to the different measurement techniques: we used mercury porosometry and the BET method whereas the authors of[11] estimated S by water absorption.

It should be noted that very high specific surfaces ($S \approx 300 \, m^2/g$) have also been reported for some microporous (non-foamed) RO systems, including urea-formaldehyde polymers[80, 81].

These high values of S for plastic foams may only be explained by the existence of a large number of microcells. Indeed, calculations of S from optical measurements (using a light microscope) by multiplying the number of macrocells by the doubled area of spherical surface gives values which are by $3-4$ orders of magnitude smaller than those obtained by measurements using absorption methods.

Therefore, oligomeric foams may be assigned to high-dispersion systems not only on the basis of formal criteria; they indeed exhibit properties of such systems. Thus, Berlin, Shutov and Aseeva showed[4, 7, 10] that phenolic foams display typical properties of high-dispersion systems: with increasing specific surface, the thermooxidative stability is reduced by several dozens of degrees. Recently, Kopshev, Shoshtaeva and Korotkov[82] have reported that the flammability of PUR foams sharply decreases with increasing apparent density (Fig. 5) with a simultaneous reduction

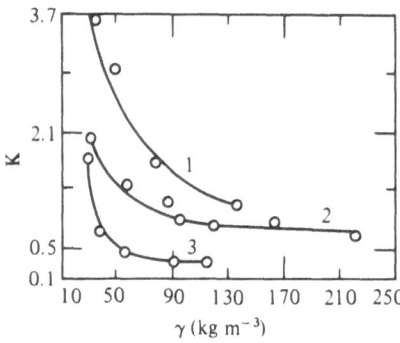

Fig. 5. Dependence of flammability index K on apparent density γ for three grades of PUR foams: PPU-18N(1), PPU-13N(2) and PPU-TS(3)[82]

of mass loss and increase of the oxygen index. However, they did not give any interpretation of these data. Taking into account the consequences of the dependence of the specific surface on the apparent density of oligomeric foams (including PUR foams), the regularities observed in Fig. 5 may be explained as follows: For each of the three groups of specimens the reduction of γ causes an increase of S and consequently of flammability. Therefore, the data obtained in[82] support our view: the primary reason for the reduction of thermal oxidation (burning) resistance is not the decrease of apparent density but the increase of specific surface. (For further properties of dispersion systems in oligomeric foams see Chap. 6.3).

In classical colloidal chemistry, solid foams were traditionally assigned to low or coarse dispersion systems, i.e. to systems with minimal pore (cell) sizes of about 1 μm or more and with a degree of dispersion (specific surface) of $10^5 - 10^6$ cm^{-1}.

From the point of view of colloidal chemistry such assignment was quite correct since the minimal size of cells did indeed considerably exceed the minimal value of

1 μm and S $\ll 10^5 - 10^6$ cm^{-1}.

However, the detection of microcells in the structure of several oligomeric foams on the one hand, and the measured values of the specific surface on the other hand, require a re-examination of our notions about the degree of dispersion in these materials.

As shown above, the minimal size of microcells is in the range of 10^{-2} μm (R $\leqslant$ 1000 Å) and S in the range of 10^5 cm^{-1}. Thus, recent data on the structure of oligomeric foams, at least those described above, allow assignment of these foams to high or fine dispersion systems.

The assignment of oligomeric foams to fine dispersion systems is of considerable importance since it enables to approach the study and modification of structure by using the concepts of dispersion media. The advantages of such an approach are evident if one takes into account the variety of ideas and methods developed in the study of high dispersion systems of "non-polymeric" nature and the first results of the application of this approach to oligomeric foams[4, 7, 9, 67].

It can be expected that this approach will result in a new physical (microstructural) rather than a chemical method of control and modification of physicochemical properties, by variation of the specific surface, the number and size of macro- and microcells. Taking into account the fundamental role of the specific surface in the processes of thermal oxidation, combustion, sorption, mass and gas transfer, one may also hope for the development of a more accurate "mathematical apparatus" for predicting the entire complex of physico-mechanical properties and of the behavior and aging in various temperature-humidity media. The development of such an apparatus should be based on quite different model notions about the morphology of oligomeric foams, accounting for both the microstructure and macrostructure.

5.5 Open Porosity

It is well known that, depending on the field of application, preference is given either to closed-cell or open-cell foams. The latter are widely used for sound and vibration insulation, in chromatography, as absorbents and filters of liquids and gases, for the preparation of metallized foamed materials, etc.

Considerable information has been accumulated regarding the relationship between the portion of open cells and the technical properties of oligomeric foams[4, 74, 83–85]. There are considerably less data available on the relationship between the amount of open cells and other morphological parameters of RO foams, for example apparent density. Reticular foams based on RO are not considered in this survey because the open-cell structure is created by secondary processing of finished products and not during foaming.

In many works it has been established that the relative number of open cells increases with decreasing apparent density of RO foams. This phenomenon has been discussed in detail in[4]: The increase in the size of cells during foaming is accompanied by a decrease in wall thickness, that finally results in a rupture of the polymeric films forming the cell membranes and in the occurrence of open, i.e. communicating cells.

It was considered not so long ago that, along with the thickness of membranes, that of cell struts (ribs) also decreases with expansion of gaseous cells. However, first Menges and Knipschild[86] and then Barma et al.[87] showed that the thickness of struts d increases with increasing cell size (length l of the rib of a pentagonal dodecahedron forming the cell). This dependence for the PUR foam is expressed as $d = l^{0.75}$ [86].

According to the data obtained by Shutov and Chaikin[88] for several RO (polyurethane and phenolic) foams, the curve $\vartheta_\alpha = f(\gamma)$ exhibits two maxima where ϑ_α is the volume fraction of open cells (Fig. 6). One of these maxima indeed corresponds to low values of γ, while the other one is in the range of γ values corresponding to the volume fraction of a gas, i.e. $\vartheta_g = 0.67–0.75$. Note that values of porosity corresponding to closest spherepackings, i.e. rhombohedral (P = 0.74) and tetragonal (P = 0.69) packing, lie in the same range of γ values. Such a coincidence cannot be considered accidental since it is known that for closest sphere packings, the number of contacts increases to $10–12$[5]. A closer packing enhances the transition of the spherical system of gas bubbles to a polyhedral system. This leads to an increase in the probability of the destruction of walls in polydisperse cells due to the difference in internal pressures of polyhedra with different numbers of faces and varying radii of wall curvature (see Chap. 6.4). On the other hand, Salyer et al.[83] reported an inversely proportional dependence $\vartheta_\alpha = f(\gamma)$ with considerably higher values of ϑ_α for PU foams (Fig. 6). However, they studied so-called precipitation foams the porosity of which is created not by foaming but by evaporation of the solvent from the formulation which is present in the form of a gel. This is why the mechanism of cell

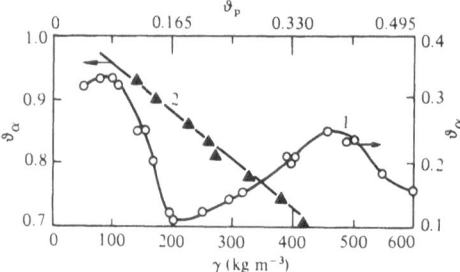

Fig. 6. Dependence of the volume fraction of open cells ϑ_a on apparent density γ and volume fraction of polymeric phase ϑ_p for PUR foams: 1: composition based on polyester foamed by carbon dioxide; 2: precipitation plastic foam, (according to data of Salyer and co-workers[83])

opening involving deformation and destruction of gas bubbles due to the packing factor cannot be applied to such foams.

A more effective use of open-cell foams requires a better understanding of the mechanism of cell opening which would ultimately enable to obtain materials with a predetermined number of open (or closed) cells.

In the early 1960's Saunders and Frisch[78] proposed a colloidal-chemical mechanism of open-cell formation in oligomeric foams. Later, Rossmy et al.[74] formulated a physical mechanism of cell opening due to the effect of water vapor (see Chap. 5.3). The data presented in this section explain the formation of open cells on the basis of "morphological" factors taking into account the type of packing of gas bubbles in oligomeric foams.

Another study of foamed polymers uses modern physicochemical methods of structure elucidation of gas-filled polymers and more accurate mathematical description. These studies have considerably contributed to the understanding of the general character of the spatial structure of plastic foams. Thus, quantitative estimations of the effect of each morphological parameter (specific gravity, size and shape of cells, type of communication between cells, cell distribution in the bulk, etc.), on the properties of a given material could be made.

A further elucidation of the nature of foamed plastics requires an even more extensive use of methods of physics and physical chemistry of polymers. In this way, a quantitative estimation of the "oligomeric" specificity of foam morphology will be possible and especially the relationship between chemical and supermolecular organizations of walls and struts in gas-structure elements and the morphological parameters of foams could be established.

A further study of the microcellular structure of oligomeric foams and the development of models taking into account both macro- and microcells are equally important.

The next stage in morphological studies should be concerned with the examination of plastic foams from the point of view of the physics and mechanics of discrete media.

Knowledge of morphology is essential for the prediction of the technical characteristics and behavior of plastic foams subjected to various external conditions.

6 Moisture and Water Absorption of Oligomeric Foams

Studies of the moisture absorption (hygroscopicity) and water absorption (hydroscopicity) of gas-filled plastics are of considerable practical importance, since foamed polymers always contain moisture (with the exception of their use in space or under extremely rare conditions on Earth) which noticeably affects all physico-mechanical characteristic of materials, in particular electrical and heat insulation properties.

It is no coincidence that moisture and water absorption are obligatory characteristics of any new material or article based on plastic foams which, together with other properties, determine the range of application. Numerous experimental data have been accumulated over the last few years on the effect of moisture on the physico-mechanical properties of oligomeric foams exposed to different temperature-moisture environmental conditions.

Nevertheless, in the field of physical chemistry of foamed polymers the "transition from quantity to quality" has not yet occurred; practically no generalizations, even of semi-empirical nature, are available which relate the kinetics of moisture and water absorption to the main morphological parameters of polymeric foams (specific gravity, portion of open cells, etc.) Very little is known about molecular mechanisms of vapor and moisture transfer taking into account the chemical and physical structure of foams.

6.1 Kinetics and Morphology

Moisture absorption by plastic foams is a slow process and attainment of equilibrium under normal conditions requires several weeks. A generalized kinetic dependence of moisture absorption obtained by Shutov and Chaikov for various types of rigid PUR (PPU-305 A grade) and epoxidephenolic (PEN-I grade) foams is given in Fig. 7[87].

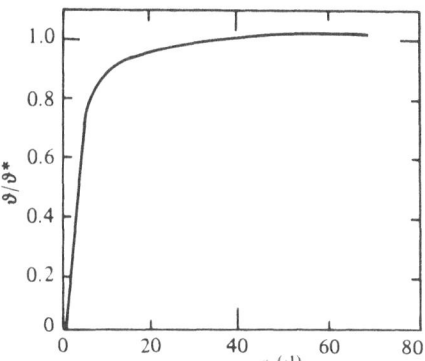

Fig. 7, Water absorption kinetics of rigid PUR foams

A statistical treatment of experimental data by the method of correlation analysis permits to suggest a simple analytical expression for the time dependence of the relative moisture absorption

$$\frac{\vartheta_w}{\vartheta_w^*} = \left(\frac{\tau}{\tau^*}\right)^{C_1}, \qquad (13)$$

where ϑ_w and ϑ_w^* = volume fractions of sorbed water and its equilibrium value; τ = current time of sorption; τ^* = time at which equilibrium moisture absorption is reached; C_1 = empirical constant ($0 < C_1 < 1$); the correlation coefficient for this dependence is 0.93, the confidence probability according to the Pirson criterion is 99.9%.

Dependences of index τ^* and constant C_1 on the chemical type of the polymer matrix and the method of surface treatment are discussed below (see Table 4).

The dependence of the equilibrium moisture absorption ϑ_w^* on morphology parameters, volume fraction of polymer (apparent density) and volume fraction of open cells is of considerable theoretical and practical interest. This dependence has been determined by Shutov and Chainkin[88] for rigid PUR (PPU-305A) and epoxide-phenolic (PEN-I) foams in the relative humidity range from 60 to 98%.

The statistical treatment of experimental data by the correlation analysis method yields the following expression:

$$\vartheta_w^* = C_2 \vartheta_\alpha^2 (\varphi)^{C_3} \exp{(-\vartheta_p/\vartheta_{max}^*)^2}, \tag{14}$$

where ϑ_w^* = volume fraction of sorbed moisture in the state of saturation; ϑ_{max}^* = maximum value of ϑ_w^*, ϑ_p = volume fraction of the polymeric phase in the foam, ϑ_α = volume fraction of open cells; φ = relative humidity, C_2 and C_3 = empirical constants ($C_2 > 0$, $C_3 > 0$) depending on the chemical type of the oligomeric matrix and method of surface treatment; the correlation coefficient is 0.9, the confidence probability according to the Pirson criterion is 99.9%.

The function $\vartheta_w^* = f(\vartheta_\alpha, \vartheta_p, P)$ (where P is the porosity of the foam) illustrated in Fig. 8 reveals that ϑ_w^* passes through a maximum at $\vartheta_p = \vartheta_p^{max}$. Note that $\vartheta_p^{max} = 0.368$ ($P = 0.632$) which, as is well known, corresponds to the closest sphere packing of cells (for the foam under consideration ϑ_p^{max} corresponds to $\gamma = 312$ kg/m³). As already mentioned, the maximum amount of open cells corresponds to the same region (Fig. 6) for the same foam.

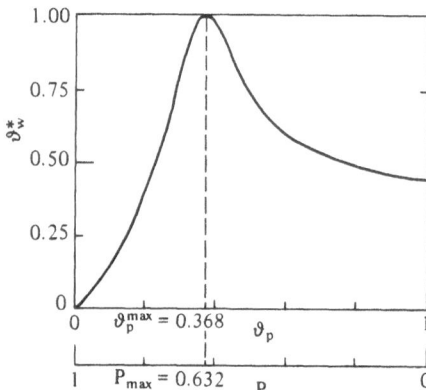

Fig. 8. Dependence of the volume fraction of saturated sorbed moisture ϑ_w^* on the volume fraction of polymeric phase ϑ_p and general porosity P for a rigid polyester-based PUR foam

In Eq. (14) the index ϑ_α is in the preexponential factor and consequently affects ϑ_w^* less than the volume fraction ϑ_p which is in the exponent. This leads to a very important conclusion: it is not the volume fraction of open cells, as usually considered, but the volume fraction of polymer and the type of packing of cells (through the value γ) which primarily determine the extent of equilibrium moisture absorption of plastic foams.

This idea has been confirmed by Lowe et al.[73] describing the dependence of water absorption of phenolic foams ($\gamma = 35$ kg/m³) on the fraction of closed cells; this dependence recalculated for the fraction of open cells is shown in Fig. 9. It is noteworthy that an increase of ϑ_α from 10 to 98% corresponds to an increase of the maximum water absorption only from 6 to 8 g/100 ml. On the basis of these data, Lowe et al. concluded that surface chemistry is more important for foamed polymers than the closed cell content. They explain the obtained data by the presence of microcells and interstices between macrocells (see Chap. 5.2).

By varying the volume fraction of polymer, i.e. the apparent density, we may, in principle, change and even predetermine the type of macrocell packing and the content of microcells (see Fig. 2), that is to affect moisture and water absorption through macro- and micromorphological parameters of the foamed polymer.

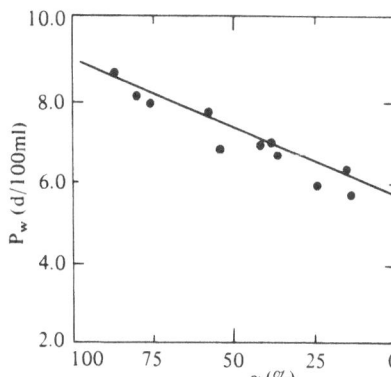

Fig. 9. Dependence of water adsorption of phenolic foams P_w on closed cell content (samples were stored under water for 7 days)[73]

We hope that further progress in the physical chemistry and technology of plastic foams based on RO will allow to realize this structural (rather than chemical) method of controlling moisture and water absorption in the near future.

Let us now return to Eqs. (13) and (14) and consider the values of τ^*, C_1, C_2, and C_3 for different cases (Table 4). The highest moisture absorption and the longest time of equilibrium attainment is displayed by samples of PPU-305 A protected by a technological skin whereas protected specimens of PEN-1 exhibit the lowest values of τ^* and C_1. This is apparently due to the different hydrophility of the polyurethane and epoxy-phenolic matrix. The fact that index τ^* practically coincides for specimens with a technological skin and without one, at the first glance, seems to indicate that Eq. (14) is incorrect, since when there are no open cells on the surface (they are closed by the skin, varnish or glass fabric), $\vartheta_\alpha = 0$ and consequently $\vartheta_\alpha^* = 0$. In practice, open microcells always exist on the surface and within oligomeric materials (including non-foamed ones) and consequently index ϑ_α is never equal to 0. This fact, among others, explains why τ^* is practically the same for protected and unprotected foams: the technological skin is permeated by a system of open microcells which, due to capillary condensation and sorption, facilitates intensive penetration of moisture into the foamed material.

Table 4. Time of saturated water absorption (τ^*) and constants C_1, C_2 and C_3 of Eqs. (13) and (14) for polyurethane (PPU-305) and epoxy-phenolic (PEN-I) foams

Surface	PPU-305A				PEN-I			
	τ^*, days	C_1	C_2	C_3	τ^*, days	C_1	C_2	C_3
Without technological skin	39.8	0.09	6.0	1.26	29.3	0.04	4.05	6.5
With technological skin	38.9	0.18	6.9	8.80	28.2	0.07	14.8	5.0
Reinforced glass fibers	32.3	0.05	4.07	2.61	36.5	0.15	1.71	1.8
Paints	32.3	0.12	2.44	2.23	29.5	0.14	4.22	1.23
Lacquers	30.4	0.23	3.30	4.30	35.1	0.05	2.84	3.7

6.2 The Sorption Mechanism

The sorption rate v of PUR foam may be calculated from sorption curves (Fig. 10). The curve $v = f(\gamma, \vartheta_p, P)$ passes through a maximum near $\gamma = 315$ kg/m^3 or $P = 0.74$ that corresponds to the closest packing of cells provided that they are spherical. The value of v decreases with the sorption time and the abscissa of the ϑ_p maximum is shifted toward higher γ values apparently indicating the beginning of capillary condensation.

The condition of capillary condensation of moisture in accordance with the Kelvin equation[89] may be expressed as

$$\ln \varphi = -\frac{2 V \sigma}{r R T} \cos \alpha, \tag{15}$$

where φ = relative humidity, σ = surface tension coefficient, V = molar volume of the liquid, τ = pore radius, R = universal gas constant, T = absolute temperature, α = contact angle between the liquid and capillary walls.

When the values of τ differ and the porous specimen is in a vapor atmosphere at fixed pressure, condensation will, according to Eq. (15), occur in pores whose

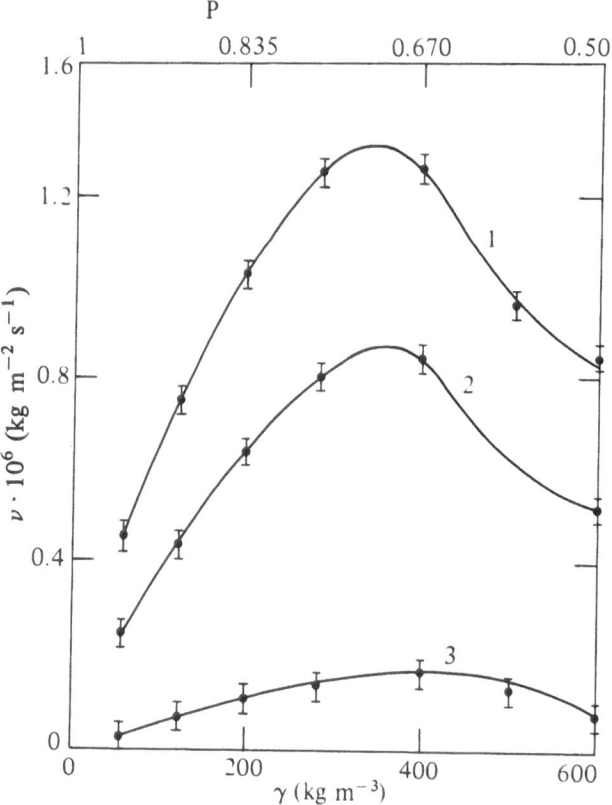

Fig. 10. Dependence of sorption rate v on apparent density γ and general porosity P at the relative humidity of 60% and at 20 °C for a rigid PUR foam; sorption times: 50 (1), 100 (2) and 1000 min (3)

radii are either equal to or smaller than τ[89]. In the range of relative humidity from 60 to 98%, τ is 25–500 Å. It follows from the data presented in Fig. 10 that about 0.5% by volume of all cells in the PUR foam and about 20% of all microcells fall within this range. Therefore, capillary condensation of moisture, being one of the main molecular mechanisms of sorption in porous materials, is quite possible during penetration of moisture into oligomeric foams and occurs due to the presence of microcells.

The specific internal surface of plastic foams, as mentioned on p. 31, may be as high as several dozens or even hundreds of square meters per gram. It is interesting to evaluate the volume fraction of the monomolecular layer of absorbed water (ϑ_w^1) on this surface. This value is calculated[12] from Eq. (16):

$$\vartheta_w^1 = S \cdot \gamma \cdot d, \tag{16}$$

where S = specific surface of the foam, γ = apparent density and d = diameter of the water molecule. Results of the calculation for a rigid PUR foam (PPU-3 grade) are given below:

γ (kg/m^3)	40	60	130	260	400	500
S (m^2/g)	73	49	22	13	8.4	7.0
ϑ_w^1 (%)	0.087	0.088	0.086	0.101	0.101	0.105

Hence, even the monomolecular layer of sorption water noticeably contributes to the total value of moisture absorption of a plastic foam.

In order to elucidate the mechanism of moisture and water absorption by phenolic foams, Lowe et al.[73] calculated the wall thickness of cells using three simplified models of cell packing: 1) spheres with point contacts; 2) dense packing of cubes without distortion of faces; 3) "fused" cubes with distorted faces.

The obtained values for two types of foams differing in the mean cell diameter are listed in Table 5. A comparison with the results obtained by scanning electron microscopy (0.5 μm) reveals that the model of dense packing of cubes is the most accurate one for fine-cell foams whereas for large-cell foams (cell diameter 400 μm) the model with fused cubes is more suitable. In[73] water vapor penetration into phenolic foams has been proposed to occur through: 1) communicating open cells: 2) microholes in walls of isolated macrocells; 3) thin films either by usual diffusion or due to intermolecular interactions during the formation of hydrogen bonds with the phenolic hydroxygroups of the resol oligomer.

Table 5. Cell wall thickness of phenolic foams[73]

Macrostructure model	Wall thickness δ (μm)	
Spheres (point-contact)	10.4	41.7
Cubes (closely packed)	5.5	21.9
Cubes (fused)	11.0	43.8
Cell length or diameter (μm)	100	400

In Chap. 5.2 we formulated the matrix model of morphology of dry oligomeric foams disregarding the third phase, water. In view of the above-mentioned data this model should be corrected as follows: under normal conditions the cellular structure of oligomeric foams is a matrix system of polymeric films covered with water films containing two groups of gaseous inclusions, sharply differing in size and relative number which are enclosed into one another.

6.3 The State of Water and Phase Transition "Water – Ice"

A detailed study of the properties and state of water absorbed by phenolic foams based on resol oligomer (FRP-3 grade) was carried out by Fedodeev and Litvinova[11] using NMR in weak fields and BET. Isotherms of distilled water absorption on foams with apparent densities from 32 to 84 kg/m^3 were recorded using a quartz balance (Fig. 11). The isotherms are S-shaped indicating polymolecular absorption: absorption terminates at $P/P_0 \geqslant 0{,}7$ (P_0 = saturated vapor pressure). From these isotherms the specific surface was calculated using the BET method (Table 6) (assuming that the landing site for a water molecule is $25 \cdot 10^{-20}$ m^2) and the curves of cell radii distribution were plotted. Data of Fedodeev and Litvinova confirm the results obtained by the author (Chap. 5.2): a large portion of the foam volume is occupied by cells with radii 200–1000 Å, and the cell size is independent of the apparent density and specific surface of the plastic foam.

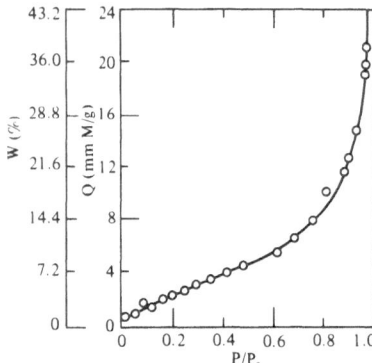

Fig. 11. Moisture sorption isotherm for phenolic foam ($\gamma = 68$ kg m^{-3}) W: moisture content by weight[11]

Table 6. Morphological and adsorption properties of phenolic foams[11]

Sample number	Apparent density γ (kg/m^3)	Specific surface S m^2/g	cm^2/cm^3	Porosity P (%)	Humidity ω (wt.%)	W (vol.%)
1	47	395	$8.4 \cdot 10^7$	94.1	94.1	18.80
2	84	321	$3.8 \cdot 10^7$	85.6	85.6	8.72
3	32	425	$1.35 \cdot 10^7$	83.7	83.7	21.84
4	68	355	$5.2 \cdot 10^7$	87.6	87.6	11.28

Because of the importance of these results we will more thoroughly examine the data obtained by Fedodeev and Litvinova. The NMR study of the state of water in phenolic foams was carried out in the temperature range from −6 to +10 °C[90]. The temperature dependence of the amplitude of free nuclear induction (Fig. 12) was treated by the least squares method; for samples 2, 3 and 4 a generalized curve was plotted on account of the considerable dispersion of the experimental points. During relaxation in a "strong" electromagnetic field (400 oersted) the temperature dependence of longitudinal relaxation time τ_1 is very weak, with a considerable scattering of experimental points. A plot of these data (τ_1 vs. T) yields a straight line with a slight slope (~ 0.01 s/deg) (Fig. 13, curve B). In "weak" field (23 oersted), on the contrary, the temperature dependence of τ_1 is more pronounced (Fig. 13, curves A).

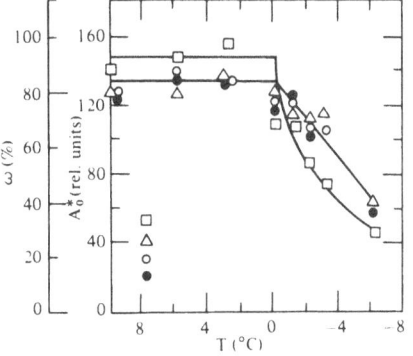

Fig. 12. Dependence of free nuclear induction amplitude A_0^* and volume content of unfrozen water ω on temperature for phenolic foams of various apparent density $\square$: 1, $\triangle$: 2, $\circ$: 3, $\bullet$: 4 (the figures denote the samples listed in Table 6)

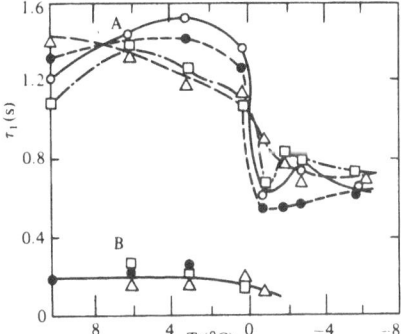

Fig. 13. Temperature dependence of longitudinal relaxation time τ_1 of water present in phenolic foams in 25 Oe (A) and 400 Oe (B) fields ($\square$, $\triangle$, $\circ$, $\bullet$: see legend to Fig. 12)[11]

Let us now consider how these data can be used for the study of the state of water in foamed plastics including the problem of phase transitions.

It follows from the data presented in Fig. 12 that at temperatures below 0 °C a strong decrease of the A_0^* amplitude is observed. This decrease is due to freezing of part of the water. For sample 1 (Table 6) with the highest porostiy the drop of A_0^* is greatest. Below −1 °C the amount of unfrozen water depends linearly (in the first approximation) on temperature; at −6 °C it is about 50% of the content at −1 °C. In accordance with a previously developed method of calculation[90] the amount of ice in a phenolic foam was determined from NMR data (Table 7).

Table 7. Ice content in (in fractions of total absorber water) phenolic foams at various temperatures[11]

Sample[a] number	Foam temperature (°C)					
	−1	−2	−3	−4	−5	−6
1	0.33	0.42	0.49	0.57	0.63	0.69
2	0.10	0.19	0.26	0.35	0.43	0.50
3	0.10	0.19	0.26	0.35	0.43	0.50
4	0.10	0.19	0.26	0.35	0.43	0.50

[a]   The samples are the same as in Table 6

The temperature dependence of τ_1 (Fig. 13) points to a noticeable effect of the nature of adsorbent (phenolic oligomer) on the properties of adsorbed water. Firstly, the value of τ_1 (1–1.2 s) is nearly half that of "free" water (in 0.5 l). Secondly, relaxation curves sharply differ in 20- and 400-oersted fields. In the 20-oersted field the dependence $\tau_1 = f(T)$ is "stepwise" and the steepest part is observed near temperatures corresponding to the phase transition "water − ice". The authors suggest that the minimum observed between 0 to −2 °C is connected with the dispersion of the relaxation time distribution. In order to confirm this assumption a "classical" relaxation analysis using deuterated water and the temperature dependence of longitudinal relaxation time is required.

The different temperature dependences of τ_1 in "weak" and "strong" magnetic fields are undoubtedly associated with the effect of the phenolic adsorbent matrix, since the weight fractions of water in the foam bulk are equal. The adsorbent effect is also revealed by the two-component character of relaxation curves (Fig. 13). Thus, the "short" component has a relaxation time of τ_1 = 0.15 − 0.20 s and is practically independent of temperature. The authors explain its existence by the presence of water in 20–100 Å large cells which is strongly bonded to the matrix[11].

Additional evidence of the different states of sorbed water in micro- and macrocells is provided by the calculation of water viscosity (η) according to the model of relaxation of molecular diffusional motion:

$$\frac{1}{\tau_1} = \frac{\gamma_1^4 h^2}{5} \left(\frac{6\pi a^3}{b^6} + 9\pi^2 N \right) \frac{\eta}{kT}, \tag{17}$$

where γ_1 = gyromagnetic ratio of the nucleus, h = Planck's constant, a = radius of water molecule, b = distance between protons, N = number of protons in a volume unit, k = Boltzmann's constant, T = absolute temperature.

It has been found that the viscosity of water in microcells (micropores) is by an order of magnitude higher than that of common "free" water and is about 10 cps in the temperature range from +10 to −6 °C (the effect of oxygen dissolved in "free" water which leads to a two-fold reduction of τ_1 was taken into account). In macrocells at 0–10 °C, the water viscosity is 2 cps and at temperatures below −4 °C, about 4 cps[11].

In order to convert the distribution of cell volumes into radii, Fedodeev and Litvinova used the well known theory of capillary condensation describing the relationship between the reduction of temperature of the water – ice phase transition (ΔT) and the radii R of pores in which this transition occurs

$$R = \frac{2\,T\,\sigma\,V}{\Delta H \cdot \Delta T}, \qquad (18)$$

σ = surface tension of water, V molar volume of water, ΔH = molar heat of phase transition (melting of water), $\Delta T = T - T_0$ (T_0 = temperature of phase transition under normal conditions).

The radii of pores in which water is frozen at -1 to $-6\,^{\circ}$C are determined from Eq. (18). The volume of water, which undergoes phase transition into ice at these temperatures, is determined from data of Tables 6 and 7; the obtained results are plotted in coordinates V vs. R. As can be seen from Fig. 14, these dependences are similar for all samples and differ only in the absolute values of V, due to the different apparent densities of samples. Thus, for sample 1 (γ = 32 kg/m^3) the specific volume of pores (R = 600 Å) is V = 2 cm^3/g, for sample 2 (γ = 84 kg/m^3) V = 1 cm^3/g, etc.

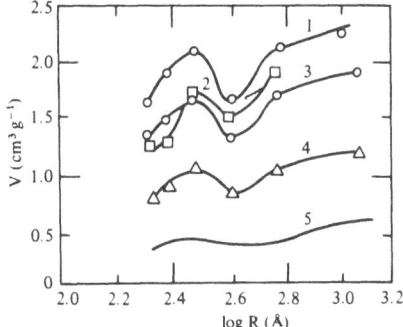

Fig. 14. Distribution curves of cell volume V in a phenolic foam according to free nuclear induction and BET data (numbers at the curves see Fig. 12 and Table 6); 5: averaged curve corresponding to sorption data[11]

In all the samples studied the specific volume of pores is inversely proportional to the apparent density, independently of the radius. Curve 5 obtained by the free nuclear induction method is given in Fig. 14 for comparison. As can be seen, it differs only in absolute values of the specific pore volume which, for absolute moistening, are lower by 2–4 times. This difference may be explained as follows. Firstly, for the recording of adsorption isotherms, samples were evacuated and then saturated by water vapor at relative pressure P/P_0 = 1. In this case, the foam is in contact with vaporized moisture and not all the cells are saturated: a portion of them are not filled because of their large size or because of the inaccessibility to water molecules of end, "bottle-like" and other forms of pores. On the other hand, in NMR studies, the foam is in direct contact with the liquid. Secondly, the phase transition "water – ice" may have a substantial effect on the character of the curves V = f (R). In large cells moisture is redistributed in the formation of crystallization sites. Growth of ice crystals the volume of which is nearly 10% larger than that of the initial water leads to a deformation of cell walls and ultimately to their destruction. Thus abrupt freezing-melting cycles in phenolic foams result in a noticeable reduction of strength. On the other hand, a "smooth" (even multiple) passing over 0°C does not affect the strength characteristics of these materials[73].

6.4 Phase Transitions at Cryogenic Temperatures

Moisture absorption of plastic foams is closely connected with cell properties, including thermal insulation properties. Investigation of heat insulation of plastic foams containing a liquid phase helps to elucidate the mechanisms of moisture and mass transfer as well as of heat transfer in gas-filled materials.

Variation of thermal conductivity coefficients of oligomeric foams at room and higher temperatures and at various degrees of humidity has extensively been studied[4, 91, 92]. Information about heat insulation at negative temperatures[34, 93, 94] and especially cryogenic temperatures (below −200 °C) is much more scare, although it is of considerable importance since oligomeric foams are frequently used in refrigerating engineering.

In one of the few studies of plastic foam behavior at cryogenic temperatures Hingst[95] discovered that upon cooling of closed-cell PUR foam in the 30–70 K range, an abrupt change in the linear temperature dependence of heat conduction occurs (Fig. 15). This phenomenon is closely related to the problem which we are

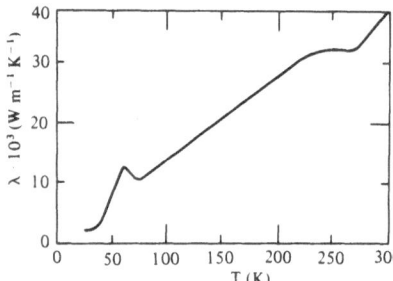

Fig. 15. Temperature dependence of the thermal conductivity, λ, of a rigid PUR foam ($\gamma = 30.8$ kg m^{-3})[95]

discussing: how and according to what mechanism does the liquid phase in a plastic foam affect the physico-mechanical properties. In the temperature range under consideration, condensation products of air components (oxygen, nitrogen, etc.) act as the liquid phase. Samples studied by Hingst also contained 30% of fluorocarbon-11 which condensates at much higher temperatures (see p. 45).

Hingst explains the abrupt change on the $\lambda = f(T)$ curve by "gas – liquid" phase transition. A further reduction of coefficient λ is caused by the phase transition "liquid – solid".

Hingst suggested a mechanism which he called "the effect of thermal tubes" to explain the increase of λ during condensation of the gaseous phase. The macrostructure is viewed as a regular polyhedral structure consisting of equidimensional dodecahedra (Fig. 16a), which is permissible for a light foam ($\gamma = 30.8$ kg/m^3). A dodecahedron consists of 12 rhombic faces of equal area arranged at an angle of 120° with respect to each other.

During phase transition (condensation), under the action of capillary forces, the liquid is collected between two rhombic surfaces forming a channel (Fig. 16b) with curved surface R_1 (curvature radius of the liquid along channel R_2 is infinite).

The value of R_1, which depends on the amount of liquid in the cell, may be calculated at known gas density γ_g before condensation, saturation density of gaseous phase γ_g' at the measuring temperature, density of liquid γ_f and cell diameter of dodecahedron D:

$$R_1 = 0.8 \left[\frac{D^2}{\gamma_f} (\gamma_g - \gamma_g') \right]^{1/2} \tag{19}$$

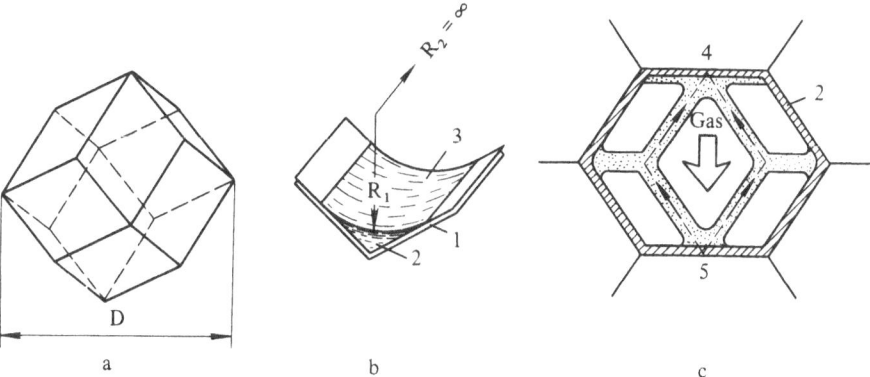

Fig. 16a–c. Dodecahedral model of cell (a); a portion of this cell containing liquid phase (b) and scheme of liquid and gas phase flow in cell (c); R_1 and R_2: curvature radii; 1: cell wall; 2: liquid phase; 3: liquid phase surface; 4: evaporation zone; 5: condensation zone[95]

Fig. 16c shows the cross section of the dodecahedral cell (channels with liquid are hatched, projections of these channels from above are denoted by dotted lines). In the plot of $\lambda = f(T)$ the upper plane of the foam sample has a higher temperature than the lower plane. Therefore, the liquid collected in the upper part of the sample evaporates and the formed vapor moves down to condensate in the lower part (Fig. 16c). As a result, the curvature radius of channels in the upper part R_1^u decreases, whereas in the lower part it increases R_1^d.

The difference in curvature radii creates a pressure drop in cell capillaries in accordance with Laplace-Young's law

$$\Delta P_c = \sigma \left(\frac{1}{R_1^u} - \frac{1}{R_1^d} \right) \tag{20}$$

The pressure drop created between capillaries in evaporation and condensation zones forces the liquid upward, so that in each dodecahedral cell circulation of liquid and gas as well as pressure drops ΔP_f and ΔP_g occur. Due to gravitation, hydrostatic pressure ΔP acts in each cell. Therefore, the suction effect of capillaries is

$$\Delta P_c = \Delta P_f + \Delta P_g + \Delta P \tag{21}$$

According to Poiseuille's law

$$\Delta P_{f,g} = \psi \, \frac{L \gamma_f V}{2 \, d_f}, \tag{22}$$

where d_f = hydraulic diameter, L = capillar length, γ_f = density of the liquid phase, v = flow rate, ψ = resistance coefficient; $\psi = 64/Re$ since the flow is laminar (Re is Reynolds number).

Since the values of ΔP_g and ΔP are negligible, in accordance with condition $\Delta P_c = \Delta P_f$, we can determine the flow of substance $Q_f = Q_g$ which corresponds to the additional flow of heat Q_{add} in the foam cell

$$Q_{add} = Q_f \cdot r_v, \tag{23}$$

where Q_f = thermal flow of the liquid and r_v = evaporation enthalpy.

The additional heat flow Q_{add} is transferred by "liquid – vapor" circulation and the contribution of this flow to the total coefficient of thermal conductivity is unrelated to the four well-known mechanisms of heat transfer in plastic foams: heat transmission of gas; heat transmission of the polymeric matrix, heat radiation and convection.

If each dodecahedron cell is considered as a miniature tube, the plastic foam consists of N such tubes arranged parallel to each other, and the total thermal flow Q_{add}^{Σ} is

$$Q_{add}^{\Sigma} = Q_{add} \cdot N = \lambda_{add} \cdot A \cdot \Delta T / a, \qquad (24)$$

where A = surface area of the foam, ΔT = temperature drop on its surfaces and a = sample thickness.

For the sample studied

$$\lambda = 4.4 \cdot 10^{-4} \frac{\sigma \cdot r_v}{\eta_f \gamma_f} \cdot \frac{a}{D} \cdot \frac{\gamma_g - \gamma_g'}{\Delta T} \cdot \frac{\Delta R}{\Theta}, \qquad (25)$$

where σ = surface tension, η_f = kinematic viscosity of the liquid phase, Θ = correction factor, $\Delta R = R_1^d - R_1^u$: difference in curvature radii of water channels in the evaporation and condensation zones.

The term $\sigma \cdot r_v / \eta_f \gamma_f$ in Eg. (25) only describes the content of the gas filler in the foam and expression a/D is a function of sample geometry. The deviation of the proposed formula from the experimental curve does not exceed 25% at $\Delta R/\Theta = = 1.5 \cdot 10^{-4}$ mm (for the system polystyrene foam – air). This relatively pronounced departure can be explained by the selection of the macrostructure model, which does not account for the dimension and shape distribution of cells, and by the inaccurate assumption that only one type of liquid is formed during condensation. Each component of air has its own condensation and freezing temperature and Eq. (25) should include separate parameters σ, η, γ, and r_v and not averaged values.

The gaseous phase of PUR foams is considered to consist only of air although it contains 30% of fluorocarbon-11. However, this does not affect the general conclusion. Indeed, if the proposed mechanism is correct, the condensation of fluorocarbon vapor should result in an abrupt change in the shape of the $\lambda = f(T)$ curve. The resultant peak is observed in the region of 270–300 K corresponding to the condensation and freezing temperatures of fluorocarbon-11 (Fig. 15).

The suggested mechanism of heat transmission and the scheme of liquid phase circulation in plastic foams[95] may be used to explain the behavior of gaseous, liquid and solid phases of water at normal and negative temperatures.

7 Electrical Properties of Oligomeric Foams

7.1 Dielectric Permeability and Dielectric Losses

Numerous experimental data on dielectric permeability ϵ and dielectric losses tg δ of oligomeric foams at various frequencies have been collected. For all plastic foams the following linear relationship exists between ϵ and γ (Fig. 17):

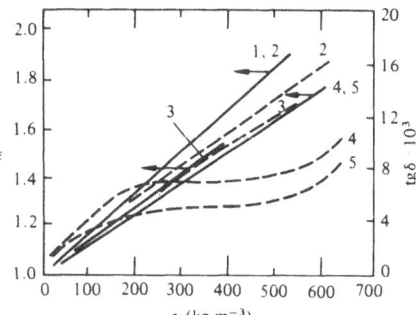

Fig. 17. Dependence of the dielectric permeability ϵ and of the tangent of dielectric loss angle tgδ on the apparent density of rigid oligomeric foams. Epoxide – PE-8 (1) and PE-9 (2); polyurethane – PPU-204 (3), PPU-305 A (4) and PPU-307 (5)[111]

$$\epsilon = 1 + K\gamma, \tag{26}$$

where K = empirical coefficient dependent on the chemical nature of the oligomeric matrix and varying from 1.3–1.5 (polyurethane and epoxide foams) to 1.6–1.8 (phenolic foams).

The disadvantage of this simple relationship is that it does not show how coefficient K depends on the frequency of dielectric dispersion (expression for the frequency dependence see p. 48). Nevertheless, Eq. (26) implies that a dry plastic foam may be considered as a laminated two-phase system of the type "dielectric – gas" with layers arranged parallel to the electric field lines. This conclusion is based on the well-known observation that a linear dependence (Eq. (26)) indicates a precise solution to the problem of dielectric permeability of a laminated dielectric with layers parallel to the field.

In the first approximation, tg δ = f (γ) is described by Borodin's formula[96]

$$\text{tg } \delta = \text{tg } \delta_p \frac{2\,\gamma}{2\,\gamma_p + \gamma} \tag{27}$$

where subscript "p" designates the polymeric matrix.

Kopatsky and Chaikin[97] proposed a formula which allows to calculate with high accuracy ($\pm 3\%$) the value of ϵ for all types of oligomeric and polymeric foams over a wide range of apparent densities

$$\epsilon = 1 + (\epsilon_p - 1)\,\vartheta_p, \tag{28}$$

where ϵ_p = dielectric permeability of the polymeric matrix and ϑ_p the volume fraction of the polymeric phase.

Equation (28), just as Eq. (26), is a solution for the matrix rather than for statistical systems or, more precisely, for a laminated dielectric with layers arranged parallel to the field.

Shutov and Chaikin[69] determined the Pirson criterion for all known formulae used for the calculation of ϵ of two-phase systems, including plastic foams. The Pirson criterion is expressed as

$$\chi^2 = \sum_{i=1}^{N} \left[\frac{\epsilon_i^e - \epsilon_i^t}{\sigma(\epsilon_i)}\right]^2, \tag{29}$$

where N = volume of sampling, σ = standard measurement error, ϵ_i^e and ϵ_i^t = experimental and theoretical (calculated) values of dielectric permeability of plastic foams.

In addition to Eq. (28), the following relationships have been examined[98, 99]:

Levin

$$\epsilon = \epsilon_p \frac{3 \gamma_p + 2 \gamma (\epsilon_p - 1)}{3 \gamma_p \epsilon_p - \gamma (\epsilon_p - 1)} \tag{30}$$

Maxwell-Winer

$$\frac{\epsilon - \epsilon_p}{\epsilon + 2 \epsilon_p} = \vartheta_g \frac{\epsilon_g - \epsilon_p}{\epsilon_g + 2 \epsilon_p} \tag{31}$$

Bruggeman

$$\frac{\epsilon - \epsilon_p}{\epsilon_g - \epsilon_p} = \vartheta_p \left(\frac{\epsilon}{\epsilon_p} \right)^{1/3} \tag{32}$$

Rayleigh

$$\epsilon = \epsilon_p \left(1 + 3 \vartheta_g \frac{\epsilon_g - \epsilon_p}{2 \epsilon_p + \epsilon_g} \right) \tag{33}$$

Lichtenecker

$$\ln \epsilon = \vartheta_p \ln \epsilon_p + \vartheta_g \ln \epsilon_g \tag{34}$$

It has been found that the confidence probability is 90% for Eq. (28), 50% for Eq. (30), 30% for Eq. (31) 10% for Eq. (32) and less than 1% for Eqs. (33) and (34).

Therefore, the dielectric permeability of plastic foams is most accurately determined by Eq. (28). This relationship thus confirms the correctness of the model presented in Chap. 5.2 (p. 26).

Using the method of complex dielectric permeability Eq. (28) may be rewritten as

$$\text{tg } \delta = \frac{\epsilon_p'}{\epsilon'} \vartheta_p \text{ tg } \delta_p \tag{35}$$

The accuracy of this expression for dry oligomeric foams is $\pm 30\%$.

The volume fraction of polymeric phase is given by $\vartheta_p = \gamma / \gamma_p$ (see Eq. (8), p. 24). Substituting Eq. (8) into Eq. (28) we obtain

$$\epsilon = 1 + \frac{\epsilon_p - 1}{\gamma_p} \gamma \tag{36}$$

A comparison of Eqs. (26) and (36) reveals the physical meaning of empirical coefficient K:

$$K = \frac{\epsilon_p - 1}{\gamma_p} \tag{37}$$

Hence, coefficient K is not simply a numerical factor as considered previously, but a frequency-dependent parameter since ϵ_p varies with frequency.

In the following the mechanism of dielectric losses in oligomeric foams will be studied. Shutov and Platonov[100] investigated the dielectric properties of a phenolic foam (FL grade) based on resol oligomer. Directly after foaming when the material still retains flexibility, the value of tg δ is very high, closer to that of conductors rather than those of dielectrics (Fig. 18). The absolute value of tg δ, however, sharply drops with increasing frequency ν. This suggests that, at the initial stages of hardening, dielectric losses in foams are due to conduction current. This is in agreement with the ionic mechanism of phenolic oligomer polycondensation[101–104]. However, if one considers a simultaneous decrease of ϵ with increasing frequency, it turns out that conduction current is not the only cause of dielectric losses in freshly foamed plastics. Increasing the frequency from 1 to 100 kHz results in a reduction of tg δ from 5.5 to 0.1 and not to 0.055 as expected for a purely ionic mechanism of conduction when the ratio 2π tg $\delta = const$ is valid[99].

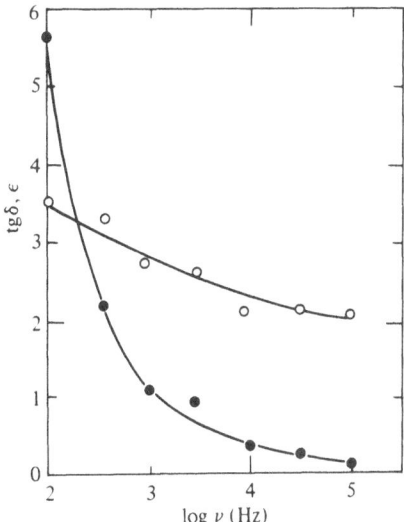

Fig. 18. Dependence of ϵ and tgδ on the frequency of the electric field for freshly foamed phenolic plastics ($\gamma = 200$ kg m^{-3})

The additional dielectric losses in phenolic foams may be attributed to the following: Firstly, polar chemical groups participate in dipolar polarization; NMR and IR spectroscopic studies[4] reveal that methylol groups act as such groups. Secondly, the restriction of charge transfer in the dielectric due to the existence of several phases in its macrostructure, leading to Maxwell-Wagner losses as a result of charge accumulation on oligomer-gas interfaces. Evidently, the contribution of these losses to the total value of dielectric losses should be particularly high in the case of plastic foams.

7.2 Electroconductivity

Electroconductivity κ is one of the most important characteristics of foams. For solid foams it determines the reliability of a foam as an electrical insulator, for liquid foams it serves as a technological parameter which is used to monitor foaming and hardening kinetics of gas-filled polymers.

A general expression for the electroconductivity of two-phase systems with a random arrangement of spherical particles of the disperse phase is given by Wagner

$$\frac{\kappa}{\kappa_m} = \frac{2(1 + \vartheta)}{2 + \vartheta} ,$$ (38)

where κ and κ_m = electroconductivity of disperse system and dispersion medium, respectively; ϑ = volume fraction of the disperse phase.

The volume of the dispersion medium (air) is related to the degree of foaming K by ratio $\vartheta = 1 - (1/K)$ and Eq. (38) may therefore be rewritten as

$$\frac{\kappa}{\kappa_m} = \frac{2}{3K - 1}$$ (39)

for systems with a high degree of foaming (■ 100) Eq. (39) coincides with Manegold's equation[105]:

$$\frac{\kappa}{\kappa_m} = \frac{2}{3K}$$ (40)

Calculations of the generalized conductivity (electroconductivity, thermal conductivity, dielectric and magnetic permeability) of heterogeneous systems have been carried out by Kerner[106] and Odelevsky[107]. For a matrix heterogeneous system with cubical inclusions whose centers form a cubic lattice and whose faces are parallel, Odelevsky's relationship may be applied:

$$G = G_m \left[1 + \frac{\vartheta}{\frac{1 - \vartheta}{2} + \frac{G_m}{G_p - G_m}} \right] ,$$ (41)

where G, G_m and G_p = generalized conductivity of disperse system, dispersion medium and disperse phase, respectively.

Provided that $G_m \gg G_p$ (for liquid foams κ of solutions $\gg \kappa$ of air) we obtain Eq. (39) from Eq. (41) by substitution of κ instead of G. In contrast to Wagner's formula, Odelevsky's formula holds for all concentrations of the disperse phase (gas) and for all types of gas-filled systems: gaseous emulsions ($\vartheta \leqslant 0.74$), spherical ($0.74 < \vartheta < 0.9$) and polyhedral ($\vartheta > 0.9$) foams. It requires isotropy of the matrix structures and equal diameters of the disperse phase inclusions. Therefore, the dependence of the ratio of the foam to the solution electroconductivity on the degree of foaming in the general form is given by equation

$$\frac{\kappa}{\kappa_m} = \frac{a}{K}$$ (42)

RO foams exhibit the highest degree of foaming among commercially produced foamed polymers. The apparent density of PF and UF foams may be as low as $3-10 \ kg/m^3$ corresponding to a degree of foaming of 300–400. Note that such light foams have not yet been obtained using high polymers. Basic physico-chemical considerations which restrict the achievement of high degrees of foaming in high polymer systems were discussed in[5].

Elucidation of the electroconductivity mechanism and accurate calculation of the electroconductivity of highly foamed RO systems is of considerable interest. Important relevant information is provided by electroconductivity studies of liquid foams with a high foaming degree, e.g. based on surfactants.

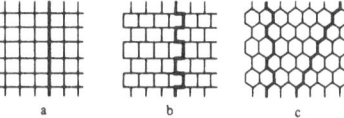

Chistyakov and Chernina[108) established that for liquid highly foamed systems based on anionic, cationic accurate.

Following the authors of[108), let us trace the current in all these schemes (Fig. 19a–c). The length of a cube rib and the side of a hexagon will be taken equal to unity and the number of rows as n. Then, the current path l (solid line in Fig. 19) in each scheme is $l_a = n$; $l_b = (3n - 1)/2$; $l_c = 2n - 1$ and for scheme 19c all current paths are equal to l_c.

The ratios

$$\frac{l_b}{l_a} = \frac{3n - 1}{3n} \text{ and } \frac{l_c}{l_b} = \frac{4n - 2}{3n - 1} \tag{43}$$

at sufficiently high n ($n \to \infty$) are

$$\frac{l_b}{l_a} = \frac{3}{2} \text{ and } \frac{l_c}{l_b} = \frac{4}{3} \tag{44}$$

For longer current routes the electroconductivity is lower and for schemes 19b and 19c

$$\left(\frac{\kappa}{\kappa_m}\right)_b = \frac{4}{9\,K} = \frac{0.444}{K} \text{ and } \left(\frac{\kappa}{\kappa_m}\right)_c = \frac{1}{3\,K} = \frac{0.333}{K} \tag{45}$$

As can be seen $a = 0.333$ is in good agreement with experimental data[108) (see above).

7.3 Dielectric Strength

RO foams are still extensively used as light dielectrics at not very high voltages in antenna caps, radar lenses, radio-transparent partitions, in panels of radio and electroequipment[4, 110). Foamed materials and non-ionogenic surfactants, the coefficient a in Eq. (42) is practically independent of solution concentration, type of surfactant and degree of foaming (K = 80–90) and is 0.35–0.39. These results do not agree with Manegold's formula which, nevertheless, is widely used for calculating the electroconductivity of liquid foams and foamed polymers.

This inconsistency has already been mentioned by Bikerman[109) who suggested that the low value of the measured electroconductivity (half the value that follows from Manegold's formula) is due to an "unfavorable" arrangement of the liquid phase.

If one imagines the structure of a liquid foam as a system of cubical gas bubbles (Fig. 19) and that the electrical current is directed upward, the horizontal walls of bubbles (perpendicular to the current direction) do not participate in electroconductivity and 2 of the 6 walls of each cube do not contribute to conduction. Then we have $\kappa/\kappa_m = 4/6\,K = 2/3\,K$, i.e. Manegold's formula is applicable. On the other hand, this equation coincides with Wagner's and Odolevsky's equations, which is to be expected since both these relations are also based on a cubical model of the disperse system.

The actual structure of highly foamed systems is polyhedral; therefore, the models proposed by Bikerman (Fig. 19b) and Chistyakov and Chernina (Fig. 19c) are more frequently used in case of high voltages, for example for lining high-voltage transformers.

According to the data obtained by Domkin[111] the value of dielectric breakdown voltage E_{br} is primarily determined by the dielectric strength of the gas in the cells and is independent of the foam density at least up to $\gamma = 600$ kg/m³ (Fig. 20).

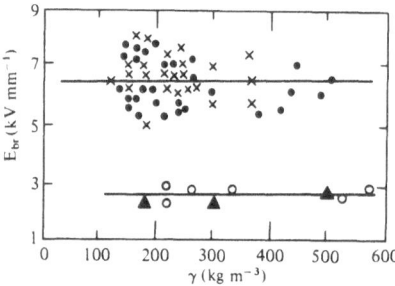

Fig. 20. Dependence of dielectric strength E_{br} on the apparent density for rigid oligomeric foams: epoxide foamed by fluorocarbon – PE-8 (x), PE-9 (•); polyurethanes foamed by carbon dioxide – PPU-307 (△), PPU-305 A (○)[111]

The technological skin at foam block edges approximately doubles E_{br} (Table 8). The higher value of E_{br} for fluorocarbons as compared with carbon dioxide and the increase of E_{br} in the presence of a technological skin have been confirmed by Palmer[112]. Findings of Kutschera and Zimmermann[113], however, are in contradiction with Domkin's work: they observed a proportional dependence between E_{br} and the PUR foam density up to $\gamma = 600$ kg/m³, although at higher γ the dependence becomes much weaker. This contradiction may only be explained by differences in the testing methods[111, 113, 114].

The basic role of gas in the electric properties of plastic foams postulated by Domkin has been confirmed by Giessner[115] in a study on the electric properties of epoxide foams, the cells of which were filled with a electronegative gas, sulfur hexafluoride (SF_6). The introduction of SF_6 results in very good dielectric properties over a wide temperature interval, including high E_{br} and corona resistance (Fig. 21). Previously, Palmer[112] established that E_{br} in the case of SF_6 is by 3.5 times higher than for air.

Therefore, the increase of the dielectric strength of foams is enhanced by the following factors:

high dielectric breakdown voltage of the gas; presence of a technological skin or density gradient in the foamed article; presence of finely separated cells; absence

Table 8. Effect of blowing agent and technological skin on the dielectric strength of PUR foams[111]

Blowing agent	E_{br} (kV/mm)	
	without technological skin	with technological skin
Carbon dioxide	2.5	4.2
Fluorocarbon-11	4.0	7.9

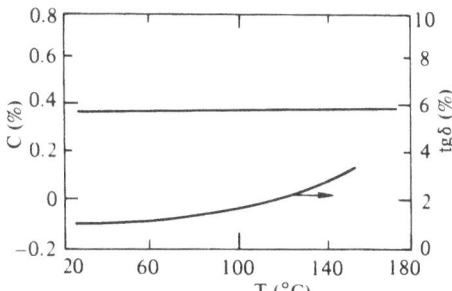

Fig. 21. Temperature dependence of SF_6 concentration C in cells and tgδ for epoxide foams[115]

of cell polydispersity; increase of foam thickness; restricted access of air and moisture to the cells.

Various additives and fillers (especially polar substances) introduced into the polymer matrix in order to increase strength, reduce flammability, etc., may drastically reduce the dielectric properties of foams.

7.4 Static Electrization

Static electrization (SE) of dielectrics increases the fire hazard of foamed constructions due to the accumulation of static electric charges and subsequent breakdown in air, resulting in "microcenters" of ignition in the form of spark dicharges. Thus, Baumann reports[57] that passage of air through open-cell UFR foams creates an electrostatic charge $q = 3 \cdot 10^{-8}$ coulomb which may cause explosions in rooms and containers with explosive gases.

Recently, attention has been drawn to methods of reducing SE of oligomeric foams by the addition of fillers (mostly metallic powders) to the initial composition[116, 117].

Shutov and Chaikin[118] measured the surface density of charges σ, created by friction for several oligomeric foams and oligomeric and polymeric monolithic plastics (Table 9). It has been found that σ of foamed plastics is by 3–4 orders of magnitude lower than that of unfoamed monolithic plastics. However, this does not imply that foams are invariably less inflammable than monolithic plastics. In three-layer sandwich composites, layer separation at the "foam-lining interface" results in additional SE. Thus, it has been shown in model tests that the removal of a glass fabric layer from the surface of PUR foam PPU-305 raises the value of σ by six orders of magnitude to $σ = 6 \cdot 10^{-5}$ coulomb/m², which is only an order of magnitude less than for air. Bearing in mind the macrorelief of foams, i.e. a possible accumulation of charges on surface protuberances, an increase of σ to breakdown values becomes quite possible.

The foregoing discussion does, however, not concern sandwich panels with metallic linings. For non-metallic linings it should be remembered that the more reliable (from the point of view of strength) the construction with a foam layer, the less inflammable it becomes. Conversely, using such a construction near ultimate mechanical strains, the imperfect technology of laminate preparation (layer separation) markedly

Table 9. Surface density σ of electrostatic charge for rigid oligomeric foams and monolithic oligomers and high polymers

Material	γ (kg/m³)	σ (coulomb/m²)
	Plastic foams	
PUR foam (PPU-3)	60–650	$2 \cdot 10^{-11}$
PUR foam (PPU-305 A)	150–400	$2 \cdot 10^{-11}$
Phenolic foam (FL-1)	70	$3 \cdot 10^{-12}$
Epoxide-phenolic foam (PEN-1)	150–400	$4 \cdot 10^{-11}$
Phenol-urethane foam (SFUP)	30–60	$6 \cdot 10^{-12}$
	Monolithic unfoamed plastics	
Polyurethane	1210	$1 \cdot 10^{-7}$
Polyethylene (film)	950	$4 \cdot 10^{-8}$
Poly(ethylene terephthalate)	2300	$7 \cdot 10^{-8}$
Poly(methyl methacrylate)	1200	$1 \cdot 10^{-7}$

increases the risk of ignition due to accumulation of electrostatic charges and subsequent breakdown of air at the "foam-lining" interface.

7.5 Dependence of Electrical Properties on Moisture

Since the electroconductivity of pure water is by several orders of magnitude higher than those of polymers, even small amounts of moisture markedly reduce electrical insulation properties of foams. For this reason, measurement of dielectric properties is a precise, rapid and non-destructive method of monitoring the kinetics and level of moisture absorption. Thus, the establishment of correlations between dielectric properties and the hygroscopicity of plastic foams makes it possible to solve two practical problems: how moisture affects dielectric properties and how to determine non-electrical properties by electrical measurements[119].

The calculation of the effect of a liquid on the dielectric properties of foams requires representation of the structure as a polymeric matrix with two-layer ellipsoidal inclusions, one of the layers being a film of sorbed moisture.

The expression for the calculation of dielectric properties of heterogeneous media containing n components was derived by Vanin[120]:

$$\epsilon - \epsilon_p = \sum_{k=1}^{n} \vartheta_k \varphi_k (\epsilon_k - \epsilon_p), \tag{46}$$

where φ_k = empirical coefficient.

For a three-component system (plastic foam) Eq. (46) may be written as

$$\epsilon - \epsilon_p = (\epsilon_w - \epsilon_p) \varphi_w \vartheta_w + (\epsilon_g - \epsilon_p) \vartheta_g \varphi_g \tag{47}$$

Shutov and Chaikin[121] showed that for rigid PUR foams

$$0.1 \leqslant \varphi_w \leqslant 0.33; \quad 1 \leqslant \varphi_g \leqslant 1.2 \tag{48}$$

For a laminated dielectric $\varphi_w = 0.1$; $\varphi_g = 1$, and Eq. (46) assumes the form

$$\epsilon = \epsilon_p + 0.1 \, (\epsilon_w - \epsilon_p) \, \vartheta_g + (\epsilon_g - \epsilon_p) \, \vartheta_g \tag{49}$$

Data calculated according to Eq. (49) (solid lines) and experimental data (points) for a rigid PUR foam at 65% relative humidity are shown in Fig. 22 (deviation $\pm 3\%$).

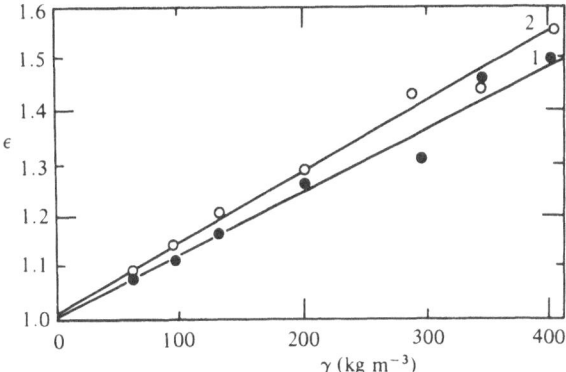

Fig. 22. Dependence of dielectric permeability on apparent density for rigid polyurethane foam (grade PPU-3) at frequencies 1 kHz (1) and 1 MHz (2) at 65% relative humidity

Using the method of dielectric permeability, Eq. (49) may be rewritten for the calculation of ϵ for plastic foams containing sorbed water:

$$\epsilon' \, \mathrm{tg} \, \delta = \epsilon_p' \, \mathrm{tg} \, \delta_p + 0.1 \, (\epsilon_w' \, \mathrm{tg} \, \delta_w - \epsilon_p' \, \mathrm{tg} \, \delta_p) \, \vartheta_w +$$
$$+ (\epsilon_g' \, \mathrm{tg} \, \delta_g - \epsilon_p' \, \mathrm{tg} \, \delta_p) \, \vartheta_g \tag{50}$$

Experimental data on rigid PUR foam coincide with 30% accuracy with values calculated from Eq. (50). It should also be noted that prolonged evacuation of PPU-305 samples containing sorbed water reduces tg δ by a factor of 20–25.

According to the data reported by Shutov and Chaikin[69] the graph of rigid PUR foam (Grade PPU-3) electroconductivity G against the volume fraction of the polymer phase displays a maximum (Fig. 23), corresponding to the maximum of moisture absorption. The analytical form of this dependence established by correlation analysis is

$$G = 10^{-12} \left[2 \, \pi^3 \exp\left(-\frac{\pi^3}{2} \cdot \frac{\vartheta_p}{\vartheta_p^{\max}} \right) + \frac{\vartheta_p^2}{\sqrt{\pi} \, \vartheta_p^{\max}} \exp\left(-\frac{\vartheta_p}{\vartheta_p^{\max 2}} \right) \right] \tag{51}$$

where G = specific volume electroconductivity in $\mathrm{ohm}^{-1} \cdot \mathrm{m}^{-1}$, ϑ_n = volume fraction of the polymer phase, $\vartheta_p^{\max} = 0.368$ (abscissa of the G maximum corresponding to the closest sphere packing of cells). The correlation coefficient is 0.983; Pirson's confidence probability = 99.9%.

For the PPU-3 foam (in the range of $\gamma = 40-500 \, \mathrm{kg/m^3}$) at 1 MHz, the variation of ϵ and tg δ with the fraction of sorbed water are given by the equations:

$$\frac{\Delta\epsilon}{\epsilon} = 3 \left(\frac{\epsilon_w - \epsilon_p}{2\,\epsilon + \epsilon_w} + \frac{\epsilon_p - 1}{2\,\epsilon + 1} \right) \Delta \vartheta_w; \quad \Delta \mathrm{tg}\delta = 3 \left(\frac{\epsilon_w \, \mathrm{tg} \, \delta_w}{2\,\epsilon + \epsilon_w} + \frac{\epsilon_p \mathrm{tg} \, \delta_p}{2\,\epsilon + 1} \right) \Delta \vartheta_w \tag{52}$$

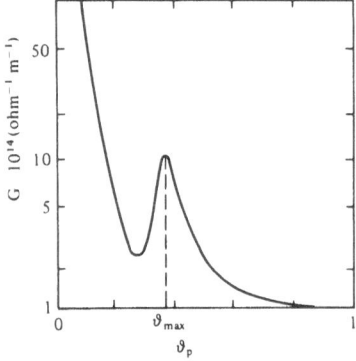

Fig. 23. Dependence of specific volume electroconductivity G on the volume fraction of the polymer-
phase ϑ_p for PUR foams ($\gamma = 40{-}500$ kg m^{-3}) in
the state of equilibrated saturation at 60% relative
humidity

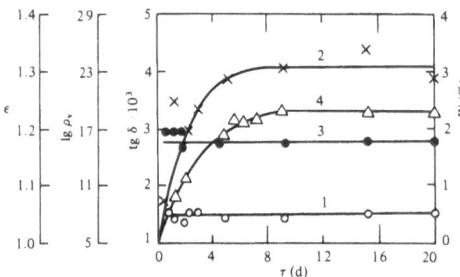

Fig. 24. Dependence of dielectric permeability ϵ (1), tangent of dielectric loss
angle (2), specific volume resistance (3)
and water absorption (4) on the curving
period for epoxide foams ($\gamma = 35$ kg m^{-3})
at 98% relative humidity [122]

Note that at 1 MHz the increments $\Delta \epsilon$ and Δ tg δ depend practically linearly on $\Delta \vartheta_w$.

The effect of moisture on the electrical properties of epoxide foams is shown in Fig. 24. These data[122] are, to a certain extent, in contrast to the above mentioned values.

Only one of the electrical characteristics (tg δ) directly depends on the moisture absorption level. According to other data[123] obtained for a semi-rigid PUR foam (grade PPU-102, $\gamma = 130$ kg/m^3), electrical characteristics are governed by temperature and humidity as follows:

Temperature (°C)	20	80	20 at $\varphi = 98\%$
tg δ ($\nu = 10^{10}$ Hz)	0.0079	0.014	0.016
ϵ ($\nu = 10^{10}$ Hz)	1.16	1.20	1.23
ρ_v (ohm · cm)	$7.5 \cdot 10^{12}$	$10 \cdot 10^{13}$	$3 \cdot 10^{13}$
E_{br} (kV/mm)	1.7	1.7	1.6

In[100] it was shown that exposure of pre-evacuated (at 10^{-3} mm Hg) hardened phenolic foam FL-1 ($\gamma = 80$ kg/m^3) to 80% relative humidity leads to equilibration of dielectric properties in 5 h (Fig. 25). The rate of sorption slows down with increasing γ (or reduction of specific surface). Thus, at $\gamma = 200$ kg/m^3 "equilibrated" values of tg δ and ϵ are reached after 3.5 h and the absolute value of tg δ increases only by 2.8 times against 4.5 for the 80 kg/m^3 sample.

Such a marked variation of dielectric properties does not occur in sorbed (or desorbed) samples of monolithic phenol oligomers. Therefore, for monolithic plastics, sorption of air moisture is a secondary phenomenon, in relation to dielectric properties, whilst for plastic foams it is a primary one.

The maxima of the curves tg $\delta = f(\nu)$ may be explained by the migration of complexes formed by polar hydroxy groups of the oligomer and water molecules involv-

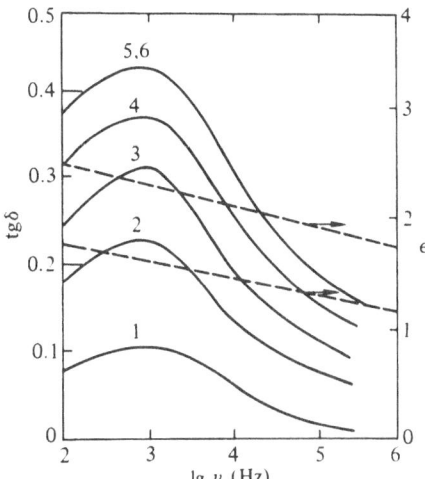

Fig. 25. Variation of ϵ and tgδ for phenolic foams (γ = 80 kg m^{-3}) dried under vacuum. 1: directly after evacuation; 2, 3, 4, 5, 6 – 1, 2, 3, 5 and 10h after evacuation, respectively

ing hydrogen bonds. It is noteworthy that with increasing tg δ, the dispersion curves do not reveal new types of dielectric losses in the hardened foam (see Chap. 7.1). This suggests a physical rather than a chemical mechanism of the phenomenon concerned. Thus, the high values of tg δ in foams are not only due to the presence of considerable amounts of water but also to a secondary phenomenon – interphase polarization resulting in additional dielectric losses, which are many times higher than in monolithic oligomeric plastics.

8 The Future of Foamed Plastics

8.1 Trends in Raw Materials and Technology

The rapidly expanding industrial production of gas filled polymers is considerably affecting the structure of both production and consumption of synthetic materials. Foamed plastics are now used practically in all branches of industry, replacing traditional natural materials (wood, glass, metals) and finding novel applications[124-128].

This rapid development is explained by a favorable combination of several technical-economical factors. In the first place, the lightness of these materials ensures a smaller consumption of polymer – along with such valuable and in many respects unique properties as buoyancy, high strength, excellent heat, electrical and acoustic insulation. In the second place, the variety of highly efficient techniques of preparation (practically utilizing all known methods of polymer processing) allows to obtain articles of any dimension and shape both in plants and directly at the place of utilization.

What are the trends in the polymeric foam industry? The production of foamed plastics on the basis of high polymers will continue to expand mainly utilizing common raw materials (polystyrene, poly(vinyl chloride), polyolefins and synthetic resins). Apart from that, one should expect a strong increase in the commercial production of foam plastics exhibiting considerably higher strength and heat durability, fire and chemical resistance, as compared with existing materials. These materials will be produced from special-purpose raw materials, aromatic polyamides and polyesters with aromatic and heterocyclic groups, polyimides, pyrones, polyarylenes, polyarylenesulfones, polyacrylates, etc.

Efforts will still be directed at improving performance properties of commercial grade foams by well established methods of modification: integration, filling, reinforcement, lamination. Along with these efforts, methods of modification based on the crosslinking of the thermoplastic matrix with various reagents, in particular reactive oligomers, will rapidly develop. The wide use of polymer — oligomer compositions will enable to broaden the variety of special-purpose materials: artificial leather, foamed fibers, foamed adhesives and sealants, artificial wood, etc.

The particular case of polymer-oligomeric compositions creates new technological possibilities such as the preparation of foamed articles directly at the place of utilization by foaming solutions or pastes at room temperature or slightly elevated temperatures. It also leads to a considerable reduction of energy consumption as compared to traditional processing methods (extrusion, casting and pressing).

In the next few years one should also expect the development of such an energy-conserving, cheap and simple method of foaming as frothing, in particular for latex systems based on copolymers of vinyl chloride, styrene and acrylates.

8.2 Reduction of Polymer Content

The introduction of foaming technology into the synthetic material industry allows to solve such an important problem as the reduction of the polymer content as compared to monolithic (unfoamed) plastics. In their wide range of application monolithic plastics, which are not subjected to considerable mechanical strains (e.g. finishing, decorative, and packaging plastics), should be developed to "slightly expanded" plastics with a density reduced by 10–30%, i.e. with a volume weight of 700–900 kg/m^3. Apart from price reduction, this allows to increase heat and sound insulating properties.

Reduction of the polymer content of foamed plastics will develop primarily in the direction of a further reduction of density. The preparation of superlight foams (including reticular ones) with a volume weight of 5–20 kg/m^3 requires special physicochemical methods since the use of conventional formulations and foaming methods of high polymers is restricted by the fact that, at small thickness of walls and struts of thermoplastic gas-structure elements the strength of the plastics is insufficient for the creation of a stable foam structure. The most efficient method of preparation of superlight foams involves crosslinking either prior to or during foaming. Crosslinking may be performed chemically (addition of peroxide or introduction of reactive oligomers), or physically (irradiation). Superlight foams will find application as thermal insulating layers in multilayer compositions, in the preparation of air and liquid filters, and as packaging materials.

The traditional method of reducing the polymer content by the addition of mineral fillers will be less used in the future since this results in a considerable increase of volume weight and complicates production. Microcapsular methods of producing "physical" foams (syntactic foams) will become more important. Reduction of the volume weight of the latter to 50–100 kg/m^3 may be achieved by employing lighter (than glass) polymeric microspheres or by forming foamed syntactic materials, i.e. by the creation of a cellular instead of a monolithic structure of the binding material. Another way of producing lighter filled foams is to use, as fillers, powders and granules from foamed materials (for example, utilizing wastes from secondary processing of plastic foams) or foamed fibers.

8.3 Further Development of Technology. Foamed Articles Instead of Foamed Materials

This section is concerned with the question of the general technological strategy in the development of new, and improvement of existing, methods for the preparation of foamed materials.

In the author's opinion the most promising route is the rapid transition from the technology of producing materials to that of producing finished products which do not require secondary treatment. It should be noted that this technique has been successfully used for a long time to prepare models from foams on the basis of reactive oligomers: cold-curing compositions enable to produce articles of the most intricate configuration directly at the site of utilization[4].

The majority of articles from high polymer foams are prepared by secondary processing of foamed materials and semifinished products (sheets, panels, blocks, etc.) by rather labour-consuming and expensive methods: glueing, cutting, fusing, machining, polishing, etc. Direct methods of producing foamed articles based on high polymers, avoiding the stage of secondary processing, have as yet insufficiently been developed.

It is noteworthy that the development of the technology of second-generation foams (early 1960's) followed this direction. For example, various methods for preparing integral foams were conceived and realized as methods of direct preparation (in one cycle only) of high-precision finished products[3].

The technical realization of the proposed route only partly depends on the replacement of existing technologies and equipment. Firstly, it requires a more profound study of all the physicochemical phenomena which govern formation, growth and stabilization of cells; secondly, elucidation of the mechanism of macro- and microstructure formation, of the effect of technological modes on the degree of foaming, morphology and physico-mechanical properties of finished products is needed, i.e. the development of scientific principles of preparation of foamed plastics.

8.4 Progress in Theory – Direct and Reverse Problems

Today, the theory of plastic foams is concerned with the solution of a problem which the author termed as a direct or physical problem. For any material including foamed materials, the direct problem is formulated as follows: in which way are the properties of finished products governed by the chemical, technological and physico-mechanical parameters of the process? However, the ultimate aim of any technology and of chemical science is, generally speaking, the preparation of substances with predetermined properties. This goal requires the solution of the reverse or chemical problem: To select, for the desired properties of a material or article not yet prepared, the chemical, technological and physico-technical parameters of its preparation. For polymeric foams such parameters include composition, method of foaming, degree of foaming, temperature and period of heating, degree of crosslinking or vulcanization, size and shape of the products, etc.

A precise solution to the reverse problem has hitherto not been obtained for any gas-filled polymer; however, approximate solutions for some parameters already exist. At present, we are able to predict the appropriate method of foaming, degree of foaming, chemical type of initial polymer on the basis of, say, given strength and thermal insulation characteristics. Less accurate is the selection of other technological parameters such as concentration and type of blowing agent, surfactant, filler, stabilizer, crosslinking agent, etc. Finally, we can only empirically determine the appropriate physico-technical parameters: absolute values and dynamics of variation of temperatures and pressures, the heating period, etc.

Results obtained by the author and included in this survey have been discussed with scientists of many countries. I am very grateful for useful discussions with my Soviet colleagues V. D. Valgin, A. M. Vasilenko, A. M. Tsukerman and I. I. Chaikin; with K. C. Frisch (USA) and T. Szvikovszky (Hungary) and, of course, to my late teacher A. A. Berlin.

9 References

1. Berlin, A. A.: Reactive oligomers and polymeric materials on their basis. First All-Union Conference on Polymerizable Oligomers, pp. 8–58. Moscow-Chernogolovka; USSR 1971. (In Russian)
2. Berlin, A. A., Matveeva, N. G.: J. Polym. Sci. Macromol. Rev. *12*, 1 (1977)
3. Berlin, A. A., Shutov, F. A.: Strengthened gas-filled plastics. Moscow, Chemistry 1980 (In Russian); Berlin, Heidelberg, New York, Springer (in translation), in preparation
4. Berlin, A. A., Shutov, F. A.: Foamed polymers based on reactive oligomers. Moscow: Khimiya 1978 (in Russian); Stamford/USA: Technomic (in translation)
5. Berlin, A. A., Shutov, F. A.: Basic principles of the chemistry and technology of gas-filled high polymers. Moscow: Nauka 1979 (in Russian)
6. Berlin, A. A.: Production of gas-filled plastics and elastomers. Moscow: GosKhimizdat 1953 (in Russian)
7. Berlin, A. A., Shutov, F. A., Aseeva, R. M.: Specific features of thermal oxidation of phenolic foams, Plasticheskie Massy *1971*, No. 12, 41 (in Russian)
8. Shutov, F. A.: On an unknown type of morphology of phenolic plastic foams. 8th All-Union Conference on High-Molecular Compounds, Kazan/USSR, 1973 (in Russian)
9. Shutov, F. A.: The gas-structure element and a new classification of plastic foams. All-Union Conference on Gas-Filled Polymers, Vladimir/USSR, 1974 (in Russian)
10. Berlin, A. A., Shutov, F. A.: Gas-structure element of plastic foams. Second Internation Conference on the Mechanics and Technology of Composite Materials. Varna, Bulgaria 1979
11. Fedodeev, V. I., Litvinova, T. A.: State of water in phenolic plastic foams. Kolloidnyi Zhurnal *38*, 756 (1976) (in Russian)

12. Shutov, F. A.: Problems and future of plastic foams based on reactive oligomers. Second All-Union Conference on the Chemistry and Physical Chemistry of Reactive Oligomers, Alma-Ata/USSR, 1979 (in Russian)

13. Sedov, A. N., Mikhailov, E. V.: Unsaturated polyester resins. Moscow: Khimiya 1977 (in Russian)

14. Berlin, A. A., Mezhikovsky, S. M.: Problems of polymer-oligomer compositions. Zh. Vses. Khim. Obshchest. 5, 531 (1976) (in Russian)

15. Berlin, A. A.: Chemistry and technology of reactive oligomers. Plaste u. Kautschuk 20, 728 (1973)

16. Plunguian, M., Cornell, E.: Process for forming foamed unsaturated polyester resins. US Pat. 3,896,060 (1974)

17. Jacobs, R. L., Backly, D. A., Simpson, J. V.: Low density resin foams. US Pat. 3,920,589 (1975), 3,920,591 (1975)

18. Minowa, S.: Porous composite materials. Japan Plastic Age 10, 36 (1972)

19. Gandini, A.: Furan derivatives in polymerization reactions. Advan. Polym. Sci. 25, 47 (1977)

20. Wade, R. C.: Flame retardant resinous foams based on furfuryl alcohol. Belgian Pat. 805,518 (1974), USA Pat. 3,865,757 (1974)

21. Van Leer Inds. Ltd.: Heat and fire resistant furfuryl alcohol foams. British. Pat. 1,487,204 (1977)

22. Van Leer Inds. Ltd.: Non-burning class 1 rating foams and a method of producing some. US Pat. 4,016,111 (1977)

23. Larsen, H. O., Barfoed, S., Cent John, A. G.: Polyfuran foams. US Pat. 3,975,318 (1976), 3,975,319 (1976)

24. VEB Farbenfab. Wolfen: Methylolketone resin foams. G. British Pat. 1,237,318 (1967), 1,238,666 (1967)

25. VEB Farbenfab. Wolten: Foamed polymers. British Pat. 1,213,116 (1966)

26. Frisch, K. C.: Miscellaneous foams. In: Plastic foams, Frisch, K. C., Saunders, J. H. (eds.), part II, pp. 770–781. New York: Marcel Dekker 1973

27. Guichard, M.: Impact resistant foams based on Coumarone indene resin. Belgian Pat. 862,226 (1978)

28. Markusch, P., Deuterich, D., Kunstler, N.: Verfahren zur Herstellung anorganisch-organischer Kunststoffe. BDR Pat. 2,559,255 (1977)

29. Lipatov, Yu. S., Sergeeva, L. M.: Synthesis and properties of interpenetrating networks. Usp. Khim. 45, 138 (1976) (in Russian)

30. Schafer, R. J. et al.: J. Cellular Plastics 14, 146 (1978)

31. Chem. Anlagen Bischo (Reut.): Foams Prepared from Phenoplast and Di- and/or polyisocyanates. BDR Pat. 2,542,900 (1977)

32. Hoechst AG (Reic.): Polyurethane foams made from oxyalkylated novolac. BDR Pat. 1,745,317 (1977)

33. Kudzio, I.: Polyurethane foams modified by urethanes. Japan Pat. 5,365,396 (1976), 5,114,0564 (1978)

34. Anonymous: New ideas in rigid foams broaden insulation capabilities. Modern Plast. Intern. 8, 50 (1978)

35. Artyushina, A. A., Tyuzneva, O. B., Chistyakov, A. M.: New casting plastic foam. Plast. Massy 1976, No. 9, 61 (in Russian)

36. Gur'ev, V. V., Sinchillo, Yu. Ya., Shutov, F. A.: Interactions of components in combined phenol – urethane foams. Plast. Massy 1978, No. 5, 73 (in Russian)

37. Berlin, A. A. et al.: Plaste u. Kautschuk 25, 697 (1978)

38. Valgin, V. D., Murasciov, Yu. S.: Materie Plastiche Elastomeru 1974, 692

39. Valgin, V. D.: Modified phenolic foams. All-Union Conf. Gas-Filled Polymers. Vladimir/USSR, 1978 (in Russian)

40. Valgin, V. D., Novak, V. A.: Phenolic foams of the type Vilares. Plast. Massy 1974, No. 10, 44 (in Russian)

41. Frank, H. G. (Rütgerswerke AG): Verfahren zur Herstellung von Schaumstoffen aus Bituminosen. BDR Pat. 1,620,847 (1977)

42. Ownes-Cornig Fiber Glass Corp.: Frothed Molding Composition. US Pat. 4,005,036 (1977)
43. Rechner Luc.: Mousse de copolymer urée formophénolique à catalyseur faible. France Pat. 7,510,368 (1976)
44. Camp de Saint Gobain: Thermally Stable Phenolic Foams. BDR Pat. 1,923,719 (1976)
45. GAF Crp.: Insoluble Porous Complexes made from Crosslinked Nitrogen USA Pat. 3,914,187 (1976)
46. Kozlov, N. A., Shorokhov, V. B.: Modification of phenolic foams. All-Union Conference on Gas-Filled Polymers, Vladimir USSR, 1978 (in Russian)
47. Kozlov, N. A., Shorokhov, V. B.: Modified phenolic foams. Plast. Massy 1979, No. 7, 55 (in Russian)
48. Mackowski, R., Majewski, S., Ostowski, K.: Phenol-polystyrene plastic foams. Polim.-Tworz. Wielkoczasteczkowe 3, 107 (1978) (in Polish)
49. Pokrovsky, V. M. et al.: Plastic foams with additives. Stroitel'nye materialy i konstruktsii 1978, No. 1, 22 (in Russian)
50. Bayer AG: Polyurethane Foams Obtained from Polyisocyanate and Polyester Containing Aminoplast. British Pat. 1,506,341 (1978)
51. Mitsubishi Gas Chem. Ind.: Modified Urethane Foam Production. Japan Pat. 52, 153,000 (1977)
52. Wagner, K. (Bayer AG): Verfahren zur Herstellung von Schaumstoffen. BDR Pat. 2,514,633 (1976)
53. Stern, G., Wegleituer, K. (Chemie Linz.): Verfahren zur Herstellung von Schaumkunststoffen. Österreich. Pat. 346,604 (1978)
54. Tokyo Rubber Ind.: Flame Retardant Acrylic Foames. Japan Pat. 7,017,876 (1977)
55. BASF AG: Magnetic Recording Elements Containing an Improved Polymeric Binder. BDR-Pat. 1,907,957 (1977)
56. Hubbard, D. A. (Imper. Chem. Ind.): Process for Producing Expanded Urea-Formaldehyde Products. British Pat. 1,463,063 (1977)
57. Baumann, H.: Preparation and Processing of Urea-Formaldehyde Foam Polymers. Jap. Plast. 10, 9, 13 (1976)
58. Balm Paints Ltd.: Amine Resins and Processes. US Pat. 4,007,142 (1977)
59. Matsushita Elect. Nor.: Foamable Flame Resistant. Japan Pat. 1,141,497 (1976)
60. Wilmsen, H.: Preparing of Flame Retardant Foamed Plastics. Swiss Pat. 588,456 (1976), 602,839 (1978)
61. Andreev, L. V., Khasanov, R. M., Prosvirin, A. A.: preparation of furan-urethane foams. All-Union Conference on Gas-Filled Polymers, Vladimir/USSR, 1978 (in Russian)
62. Seryakov, G. V., Sokolov, G. M., Voskresensky, V. A.: Casting epoxide-phenolic foam. All-Union Conference on Gas-Filled Polymers, Vladimir/USSR, 1978 (in Russian)
63. Schuur, G.: Plastica 31, 97 (1978)
64. Baumann, H.: Formaldehyde in UF-shuim. Plastica 30, 72 (1977)
65. Kozlov, K. V.: On the Determination of Coefficient of Uniformity in Foamed Plastics. Zavod. Lab. NII. 1973, 1396 (in Russian)
66. Spektor, F. A., Shutov, F. A.: Microstructure of phenolic foams. In: Physics of building materials. Oborin, L. A. (ed.), pp. 11–14. Leningrad: Building Institute 1970. (in Russian)
67. Shutov, F. A.: New approach to the study of oligomeric foam properties. All-Union Conference on Gas-Filled Polymers, Vladimir/USSR, 1978 (in Russian)
68. Shutov, F. A., Chaikin, I. I.: Micro- and macrocellular structure of polyurethane foams. Proizvodstvo i Pererabotka Plasticheskikh Mass i Sinteticheskikh Smol 1979, No. 4, 14 (in Russian
69. Shutov, F. A., Chaikin, I. I.: Electrical properties of polyurethane foams. All-Union Conference on Polyurethanes, Vladimir/USSR, 1979 (in Russian)
70. Shutov, F. A.: Problems of the physical chemistry of polyurethane foams, ibid.
71. Shutov, F. A., Chaikin, I. I.: Morphology of polyurethane foams. Fifth Internat. Conf. Cellular and Non-Cellular Urethanes, Strasburg 1980
72. Aleksandrov, A. Ya., Borodin, M. Ya., Pavlov, V. V.: Constructions with plastic foam fillers. Moscow, Mashinostroenie 1972 (in Russian)

73. Lowe, A. J. et al.: The phenol formaldehyde foams. 4./SPI Internat. Cellular Plastics Conf., Montreal, 1976
74. Rossmy, G. R. et al.: J. Cell. Plast. *13*, 26 (1977)
75. Oween, M. J., Denis, C.: J. Cell. Plast. *13*, 264 (1977)
76. Dubyaga, E. G., Tarakanov, O. G.: Khim. Prom. *8*, 607 (1976)
77. Lemlich, R.: Ind. Eng. Chem. Fundam. *17*, 89 (1978)
78. Saunders, J. H., Frisch, K. C.: Polyurethanes. New York, London: Interscience, 1964
79. Schauver, A., Truxa, K., Spitzer, Z.: The study of structure of cellular materials. Stavebnicky Casopis *15*, 245 (1967) (in Czech)
80. Salyer, I. O., Usmani, A. M.: J. Appl. Polym. Sci. *23*, 381 (1979)
81. Renner, A.: Kondensationspolymere aus Harnstoff und Formaldehyd mit großer spezifischer Oberfläche. Makromol. Chem. *149*, 1 (1971)
82. Kopshev, E. Yu., Shoshtaeva, M. V., Korotkov, L. I.: Effect of apparent density on flammability of rigid polyurethane foams. Plast. Massy *1979*, No. 2, 51 (in Russian)
83. Salyer, I. O. et al.: J. Cell. Plast. *9*, 25 (1973)
84. Mendelson, M. A. et al.: J. Appl. Polym. Sci. *23*, 325 (1979)
85. Sarig, G., Little, R. W., Segerlind, L. J.: J. Appl. Polym. Sci. *22*, 419 (1978)
86. Menges, G., Knipschild, F.: Polym. Eng. Sci. *15*, 623 (1975)
87. Barma, P. et al.: Amer. Chem. Soc., Polym. Prepr. *19*, 698 (1978)
88. Shutov, F. A., Chaikin, I. I.: Extreme character of sorptional water absorption by rigid polyurethane foams. Proizvodstvo i pererabotka plasticheskikh mass i sinteticheskikh smol *1979*, No. 6, 17 (in Russian)
89. Gregg, S. J., Sing, K. S. W.: Adsorption surface area and porosity. London, New York: Academic Press 1967
90. Fedodeev, V. I.: Estimation of properties and state of water in large-pore disperse substances. Kolloid. Zh. *37*, 520 (1975) (in Russian)
91. Hedlin, C. P.: J. Cell. Plast. *13*, 313 (1977)
92. Kirby, D.: Plast. Rubber Int. *3*, 167 (1978)
93. Kudryacheva, G. M., Kozhevnikov, I. G.: Thermophysical characteristics of plastic foams at 90–360 K. Plast. Massy *1974*, No. 5, 39 (in Russian)
94. Barnatt, A., Dyke, R., Hillier, K.: Plastica *32*, 14 (1979)
95. Hingst, U.: Forsch. Ingenieur Wiss. *43*, 185 (1977)
96. Borodin, M. Ya.: Electrical properties of plastic foams. In: Plastic foams. Popov, V. A. (ed.), pp. 61–72. Moscow: Oborongiz 1959 (In Russian)
97. Kopatsky, N. A., Chaikin, I. I.: Theoretical formula for calculating dielectric permeability of plastic foams. All-Union Conf. Physics of Dielectrics, Leningrad USSR, 1973 (in Russian)
98. Brown, W, F.: Dielectrics. Berlin, Göttingen, Heidelberg: Springer 1956
99. Sazhin, B. I.: Electrical properties of polymers. Leningrad: Khimiya 1970 (In Russian)
100. Shutov, F. A., Platonov, M. P.: On the mechanism of dielectric losses of gas-filled polymers in connection with specific features of their macrostructure. All-Union Conf. Physics of Dielectrics, Leningrad 1973 (In Russian)
101. Klemper, D., Karasz, F. E.: J. Elastoplast. *4*, 180 (1972)
102. Trostyanskaya, E. B., Chernikova, O. D.: Dielectric Properties of Phenolic and Epoxide Resins in Media with High Humidity. Plast. Massy *1976*, No. 2, 64 (in Russian)
103. Warfield, R. W.: J. Appl. Polym. Sci. *19*, 1205 (1975)
104. Harwood, C., Wostenholm, G. H., Yates, B.: J. Polym. Sci. Phys. Ed. *16*, 759 (1978)
105. Manegold, E.: Schaum. Heidelberg: Straßenbau Chem. und Techn. Verlag, 1953
106. Kerner, E. H.: Proc. Phys. Soc. *59*, 802 (1956)
107. Odelevsky, V. N.: Theory of generalized conduction of heterogeneous media. Zh. Tekh. Fiz. *21*, 667 (1951) (in Russian)
108. Chistaykov, B. E., Chernina, V. N.: Electroconductivity of liquid foams. Kolloid. Z. *39*, 1005 (1977) (in Russian)
109. Bikerman, J. J.: Foams. Berlin, Heidelberg, New York: Springer 1973
110. Anonymous: Urethane foams in the TV Receiver. J. Cell. Plast. *13*, 290 (1977)
111. Domkin, V. S.: Electrical properties of gas-filled polymers. All-Union Seminar on Plastic Foams, Leningrad USSR, 1975 (in Russian)

112. Palmer, P. J.: J. Cell. Plast. *9*, 182 (1973)
113. Kutschers, K., Zimmermann, P.: Langzeitdurchschlag in Polyurethan Schaumstoffe. Wiss.-Techn. Mitt. Inst, „Prüffeld Elek. Hochleistungstech." *18*, 28 (1977)
114. Domkin, V. S.: Electrical properties of plastic foams and methods of estimation. In: Methods of physico-mechanical testing of plastic foams, pp. 48–53. Moscow: NIITEKhIM 1976, (in Russian)
115. Giessner, B. G.: Epoxy foam – a novel electrical insulation material. pp. 336–339. Proc. 13th Electrical/ Electronics Insulation Conf., New York, 1977
116. Zorll, U.: Organische Beschichtungen zur Oberflächenreinhaltung. Ind.-Anz. *100*, 30 (1978)
117. Toyo Rubber Chem.: Antistatic Polyurethane Foam Preparation. Japan Pat. 77,035,399 (1977)
118. Chaikin, I. I., Shutov, F. A.: Static electrization of polyurethane foams. All-Union Conf. Polyurethanes, Vladimir/USSR, 1979 (in Russian)
119. Shutov, F. A.: On the use of electrical insulation from phenolic foams in high humidity media. All-Union Seminar on Plastic Foams, Leningrad 1975 (in Russian)
120. Vanin, B. V.: Calculation of dielectric properties of disperse materials. Elektrichestvo *1965*, 54 (in Russian)
121. Chaikin, I. I., Shutov, F. A.: Electroconductivity and static electrization of plastic foams. Proizvodstvo i pererabotka plasticheskikh mass i sinteticheskikh smol *1979*, No. 7, 12 (in Russian
122. Sokolov, V. A. et al.: Electrical Properties of Epoxide Foams. Plast. Massy *1977*, No. 2, 68 (in Russian)
123. Durasova, T. F., Moiseev, A. A.: Electrical properties of plastic foams at various humidity. Plast. Massy *1971*, No. 9, 19 (in Russian)
124. Bono, P., Gatland, C.: Frontiers of Space. New York: MacMillan Co. 1973
125. Larkins, C. D.: Southern Africa *7*, 41 (1977)
126. Pokrovsky, L. I.: Production and use of foamed polymers. Plast. Massy *1978*, No. 4, 68 (in Russian)
127. Kiya-Oglu, N. V., Tsokolaeva, N. M., Stanovskaya, E. R.: Trends in production and consumption of Plastic Foams. All-Union Conf. Polyurethanes, Vladimir/USSR, 1979 (in Russian)
128. Tarakanov, O. G.: Specific features of the physical chemistry and mechanics of polyurethane foams. All Union Conference on Polyurethanes, Vladimir/USSR, 1979 (in Russian)

Received February 13, 1980
G. and S. Olivé (editors)

Applications of Soluble Polymeric Supports

Kurt Geckeler[1], V. N. Rajasekharan Pillai[2,3], and Manfred Mutter[2]

[1] Institut für Organische Chemie, Universität Tübingen, D-7400 Tübingen
[2] Institut für Organische Chemie, Universität Mainz, D-6500 Mainz
[3] On leave from the Department of Chemistry, University of Calicut, Kerala, 673635 – India

Table of Contents

1 Introduction

In addition to their classical use as ion exchangers, insoluble polymers have been widely used in recent years as supports for solid-phase synthesis[1], chromato-graphy[2] and as carriers of organic reagents[3], catalysts[4], enzymes[5], and pharma-cologically active compounds[6]. The advantages of this approach are the simulation of high dilution or pseudodilution conditions, the fish-hook and concentration principle, the facilitation of selective intrapolymeric reactions, stabilization of re-active species and the elimination of volatile malodorous reagents[7]. The easy sepa-ration of the by-products of the reaction is an important advantage of the use of insoluble polymeric reagents. The question of the achievement of the site isolation and the chemical and kinetical non-equivalence of the reaction sites can be visu-alized as serious disadvantages of the reactions in such polymeric matrices[8]. In this context, the use of linear soluble polymers in the place of the cross-linked in-soluble resins would be advantageous in many respects. The separation of the poly-mers may raise an operational problem in some cases; however, in most cases, the soluble polymer permits separation by selective precipitation or by membrane fil-tration. These considerations have awakened much interest nowadays in the appli-cation of linear soluble polymers as supports for sequential type organic syntheses, as carriers of organic reagents, enzymes, catalysts, as complexing agents and as poly-meric drugs. A critical survey of the literature concerning the use of soluble poly-mers for these purposes is attempted here.

2 Applications in Organic Synthesis

Insoluble polymeric supports have been widely used during the last two decades for the stepwise synthesis of polypeptides, polynucleotides and polysaccharides. The advantages of polymer-supported reactions were fully recognized only after the introduction of this solid-phase method by Merrifield[1], although synthetic ion ex-change resins have long been used as catalysts in organic reactions. Since the insolu-bility of the polymeric reagents allows a very facile separation from the reaction mixture, a large excess of either the low molecular or polymeric reagent can be used, thereby increasing the reaction rates and yields. Polymeric reagents are also of ad-vantage in many cases if the corresponding low molecular reagent has an extreme odour or high toxicity. The nature of the polymer backbone, as the nature of the substituents, polarity etc., imparts some unique specificities to the reaction. There-fore, by suitably varying these parameters, specific, selective polymeric reagents can be designed[9, 10].

Most of the polymeric reagents which have been developed so far make use of an insoluble cross-linked polymer as the backbone. Investigations on the reaction rates and kinetic course in solid-phase synthesis revealed that the reaction sites within the polymeric matrix are chemically and kinetically not equivalent, making quantitative conversions almost impossible. Furthermore, the difficulty in the prep-aration and accessibility of insoluble polymeric reagents appears to limit a more general application of these compounds. Due to these inherent difficulties of the solid-phase method, there has recently been an upsurge of interest in the application of soluble polymers. The use of these soluble carriers permits working in a homogenous phase and all the advantages of the polymer supports are in principle fulfilled.

2.1 Linear Soluble Polymers

A polymer compatible for use in homogeneous lipuid-phase synthesis must be soluble in a variety of organic solvents and must allow the effective and complete separation of the low molecular weight components not covalently attached to the functional groups. Polymers and their functionalized derivatives which have been used so far in sequential type organic synthesis and as carriers of organic reagents are given in Table 1.

Linear polystyrene can be functionalized by various methods[10]. The functional group capacity in these polymers should not be too high; otherwise, steric complications may arise. Poly(ethylene glycol) has been found to be most suitable for liquid-phase synthesis. This linear polyether and the block copolymers with functional groups at defined distances are chemically stable and soluble in a large number of solvents including water and can be precipitated selectively. Partially hydrolyzed poly(vinylpyrrolidone) and its copolymers with vinyl acetate were successfully applied in peptide synthesis. Poly(acrylic acid), poly(vinyl alcohol), and poly-(ethylenimine) are less suitable for the sequential type synthesis because of the

Table 1. Soluble polymers and their functionalized derivatives for organic synthesis

Polymer	Functionalized derivatives/ Polymer-bound reagents
Poly(ethylene glycol) (PEG)	OH, Cl, NH_2, COOH/carbodiimide, active esters, 1-hydroxybenzotriazole, Wittig reagents
Block copolymer of PEG with OCN–R–NCO $\quad\mid$ $\quad$ X	$X = NH_2$, CH_2Cl
Poly(vinylpyrrolidone) (PVP) (15% hydrolyzed)	COOH, –NH–
Polystyrene	CH_2Cl, CH_2OH, CH_2NH_2, CH_2Li
Poly(acrylic acid)	COOH
Poly(vinyl alcohol)	OH
Poly(ethylenimine)	NH
Poly(vinylbenzoyl chloride)	COCl
Poly(vinylpyridin borane)	BH_3
Nylon 6,6	$CO-CCl_3$, $CO-CF_3$, active chlorine

limited solubility. Poly(ethylene glycol), polyvinylpyridine, polystyrene and substituted polyamides find use as carriers of various reagents for organic synthetic
purposes.

2.2 Nature of the Reagent or Functional Group Bound to the Soluble Polymer

Flory's studies[11] on the kinetics of polyesterification demonstrated that the reactivity of a functional group is independent of the size of the polymer molecules to
which they are attached. The results of the kinetic analysis of the aminolysis of
activated amino acids with high and low molecular weight esters of amino acids
were also in conformity with this general principle[12]. The rate constants of the
reaction of BOC-Gly-ONP (BOC = t-butyloxycarbonyl, ONP = p-nitrophenylester)
with poly(ethylene glycol) bound amino acids and with their low molecular weight
analogues are given in Table 2. The rates for the PEG-bound species are found to
be the same as for the analogous low molecular weight species. These results suggest that such soluble polymeric reagents follow the kinetics of the classical homogeneous system.

   The above observations with poly(ethylene glycol) may not hold for polymers
with randomly distributed functional groups of high capacity where neighbouring
effects become operative[13]. Thus, when quantitative reactions in polymer-mediated
synthesis are important for synthetic strategy, all reaction sites must be in equivalent
surroundings[8]. This concept is best accomplished in poly(ethylene glycol) and
similar derivatives with equivalent functional groups. The accomplishment of the
equivalence of the reaction sites can be regarded as the most notable advantage of
the use of these soluble polymeric supports over solid supports in organic synthesis.

2.3 Repetitive Sequential-Type Synthesis

2.3.1 Peptide Synthesis

Shemyakin et al.[14] tried to eliminate the steric complications encountered in solid-
phase peptide synthesis by using linear, non-crosslinked polystyrene of molecular
weight 200000 as a soluble support for the growing peptide chain. All coupling and

Table 2. Rate constants for the coupling reaction of BOC-Gly-ONP with different glycine esters[12]

Glycine ester	Rate constant $k_2 \times 10^{-1} (1 \cdot mol^{-1} s^{-1})$
Gly-O-Bu^t	0.23
Gly-O-Et	0.12
Gly-OPEG (MW 20000)	0.12
Gly-OPEG (MW 6000)	0.13
Gly-OPEG (MW 2000)	0.16

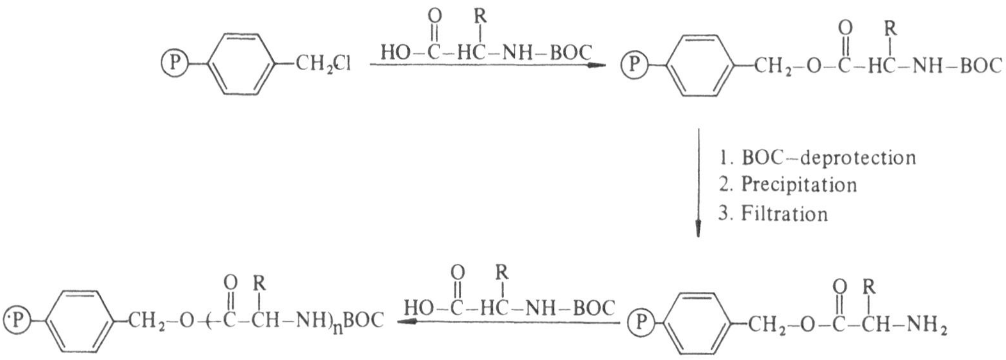

$$\text{(P)}\!-\!\!\bigcirc\!\!-\!CH_2Cl \quad\xrightarrow{\;HO-\overset{O}{\overset{\|}{C}}-\overset{R}{\overset{|}{HC}}-NH-BOC\;}\quad \text{(P)}\!-\!\!\bigcirc\!\!-\!CH_2-O-\overset{O}{\overset{\|}{C}}-\overset{R}{\overset{|}{HC}}-NH-BOC$$

1. BOC–deprotection
2. Precipitation
3. Filtration

$$\text{(P)}\!-\!\!\bigcirc\!\!-\!CH_2-O\!+\!\overset{O}{\overset{\|}{C}}-\overset{R}{\overset{|}{CH}}-NH)_nBOC \quad\xleftarrow{\;HO-\overset{O}{\overset{\|}{C}}-\overset{R}{\overset{|}{HC}}-NH-BOC\;}\quad \text{(P)}\!-\!\!\bigcirc\!\!-\!CH_2-O-\overset{O}{\overset{\|}{C}}-\overset{R}{\overset{|}{CH}}-NH_2$$

Scheme 1

deprotection reactions are carried out in solution; for the removal of the excess reagent the polymer-bound peptide is precipitated in water and washed.

This polymer-mediated stepwise synthesis in solution was claimed to be superior to solid-phase synthesis because of the higher chances of achieving quantitative coupling reactions. However, investigations of Green and Garson[15] revealed that the use of soluble polystyrene creates new sources of preparative difficulties. Most critical among these is the danger of cross-linking during the various synthetic steps. This is mainly due to the residual chloromethyl groups after the esterification step. Another serious difficulty is that the growth of the peptide chains on the polystyrene support influences its physical properties in an undesirable way, owing to the poor solubilizing power of the polymer chain upon the peptide. However, the synthesis of two model peptides by Green and Garson[15] has demonstrated that this soluble polymer technique has distinct advantages over the heterogeneous method.

Andreatta and Rink[13] used linear polystyrene of molecular weight 20400 for the synthesis of the hexapeptide Val-Tyr-Val-His-Pro-Phe. Separation of the excess reagent from the polymer-bound peptide was performed by gel filtration. The maximum capacity for maintaining satisfactory solubility properties of the peptide-polystyrene ester was about 0.5 m mol per gram of the polymer. Kinetic investigations were carried out in order to delineate the differences in the reaction rates between low and high molecular weight peptide esters[13]. A result of the major impact of the use of polyfunctional supports in general was the finding that the functional groups attached to the polystyrene showed considerable differences in chemical reactivity and kinetic behaviour. For example, in the hydrogenolytic cleavage of the benzyl ester groups joining the peptide with the polymer support, some anchoring groups proved to be totally resistant. The kinetic course of the coupling reaction also showed significant deviations from linearity due to the non-equivalence of functional groups. The enhanced reaction rates of some specific groups appear to originate from favourable polymer effects, for example, an increase of the effective dielectric constants at the reaction site[16]. A comparison of the second-order rate-constants of the aminolysis of N-protected active esters with low- and high-molecular weight amino components (Table 3) shows that the reaction rates of the soluble polymer ester are lower by a factor of about two compared to low molecular weight esters, but considerably higher than those of the heterogeneous system.

Table 3. Second-order rate constants of the aminolysis of Z-Ala-ONP with H-Pro-O-CH$_2$-R

R	k_2 $(l \cdot mol^{-1} \, s^{-1})$
C$_6$H$_6$	0.081
Polystyrene (linear)	0.047
Copoly(styrene-m-DVP)	0.037

Although the formation of truncated sequences due to sterically inaccessible reaction sites is unlikely in the homogeneous system, the difference in the chemical reactivity of functional groups, as delineated in these studies, may have a severe impact on the realization of quantitative coupling yields. The random distribution of functional sites along the linear chain molecule must be considered as a possible origin of the heterogeneous kinetic behaviour of polyfunctional supports. Thus, functional groups located at the chain ends might be less influenced by local effects of the polymer chain than groups near the centre of the chain due to differences in the segmental flexibility and in the density of the coiled chain molecule. Neighbouring effects are also to be taken into account, even when the degree of functionalization is relatively low.

From these studies it can be concluded that the intrinsic problems of the heterogeneous solid-phase synthesis, particularly the steric effects of the polymer matrix, are not solved satisfactorily by using non-crosslinked polystyrenes as support. The equivalence of all functional groups attached to a linear molecule appears to be an absolute prerequisite for the realization of homogeneous conditions which are analogous to low molecular weight systems.

Blecher and Pfaender used polyethylenimine (PEI) of molecular weight 30000 as a water-soluble support for the synthesis of a model tetrapeptide[17]. This polymer is designed for use in combination with the N-carboxyanhydride coupling method. The ultrafiltration method was found to be very effective for the removal of the excess reagent. An enzymatic method was employed in this case for the splitting of the peptide from the polymer support.

Copolymers of N-vinylpyrrolidinone and vinyl acetate were used by Bayer and Geckeler for the synthesis of peptides[18]. With symmetric anhydrides as coupling reagents several peptides could by synthesized on these supports in overall yields of ca. 60%. Owing to the presence of the pyrrolidinone ring, these copolymers exert a very strong solubilizing effect upon the peptide chain in a variety of solvents including water. However, the same synthesis using poly(acrylic acid) and poly(vinylalcohol) as supports gave less satisfactory results, mainly due to the poor solubility of the polymers. None of these polymers has found broad application in liquid-phase synthesis; from the kinetic point of view, the same objections with respect to the equivalence of functional groups are operative here as in the case of polystyrene. Yet, these deficiencies may not be so serious when considering the synthesis of small- to medium-sized peptides.

The above drawbacks are to a greater extent overcome when using linear polyethers as soluble supports for the sequential type synthesis. The presence of a hydrophilic and hydrophobic moiety per monomer unit lends this class of polymers

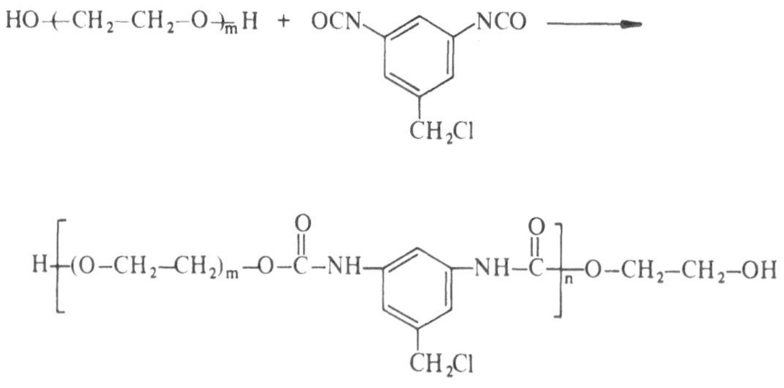

$$HO\text{-}(\!-CH_2\text{-}CH_2\text{-}O\!-)_{\overline{m}}H \ + \ OCN\text{-}\!\!\!\!\!\bigcirc\!\!\!\!\!\text{-}NCO \longrightarrow$$
$$\overset{|}{CH_2Cl}$$

$$H\text{-}\!\!\left[(O\text{-}CH_2\text{-}CH_2)_m\text{-}O\text{-}\overset{\overset{O}{\|}}{C}\text{-}NH\text{-}\!\!\!\!\!\bigcirc\!\!\!\!\!\text{-}NH\text{-}\overset{\overset{O}{\|}}{C}\right]_n\!\!\!O\text{-}CH_2\text{-}CH_2\text{-}OH$$
$$\overset{|}{CH_2Cl}$$

Scheme 2

favourable physical and chemical properties that permit their use as efficient supports for the sequential type liquid-phase synthesis of peptides, nucleic acids and
polysaccharides[19-21]. So far, mainly poly(ethylene glycol)s with molecular weights
in the range of 2000–20000 were used as the C-terminal protecting groups[22]. Recently, monofunctional polyethers were used instead of the bifunctional polyethers.
In order to obtain higher capacities of polyethers, block copolymers were synthesised
using poly(ethylene glycol) blocks and diisocyanate derivatives (Scheme 2)[23].

When poly(ethylene glycol) of molecular weight 1000 was used as the starting
material for this polyaddition reaction, polymers of molecular weight more than
20000 were obtained. As the functional groups are located in equidistant order,
neighbouring effects can be excluded in this copolymer.

The liquid-phase method for peptide synthesis takes advantage of the efficient
concept of stepwise synthesis without intermediate purification procedures. The C-
terminal amino acid is covalently attached to the hydroxy end of the poly(ethylene
glycol) which determines the physical and chemical properties of the growing peptide
chain during all stages of the synthesis. As a consequence, the polymer-bound peptide
can be readily separated from low molecular weight reagents so that the stepwise
incorporation of the amino acid residues is as efficient as in solid-phase synthesis.
However, all reactions are carried out in homogeneous solution as in classical procedures, eliminating the inherent problems of the heterogeneous reactions. Compared
to classical stepwise strategies, the use of this polymeric protecting group offers two
principal advantages:
– the solubility of the peptide is strongly enhanced by the polymeric ester group
  so that peptides with poor solubility become accessible to the stepwise strategy.
– the synthesis cycle is simplified and can be performed according to a standard
  procedure independent of the physicochemical properties of the peptide.

For an efficient synthesis using the liquid-phase method, the crystallization tendency and the solubility of poly(ethylene glycol) in various organic solvents must be
retained after attaching a peptide to its chain ends. As expected, the influence of the
peptide on these properties of the polymeric ester group depends on the primary
sequence, side chain protection, chain length, and conformation of the growing peptide chain. For example, the degree of crystallization of poly(ethylene glycol) as
determined by X-ray diffraction decreased from 80% to about 60% when the homo-
oligopeptide (Ala)$_{10}$ was bound to the chain termini of the polymer. The high reten-

tion of crystalline phases in poly(ethylene glycol) after the attachment of amorphous peptide blocks is explained by a two-phase model[24, 25].

This finding is of practical relevance for liquid-phase strategy because the partially crystalline structure of peptide-PEG esters permits easy handling of the precipitated product. Furthermore, the danger of inclusions of low molecular weight components as observed in amorphous precipitates is considerably reduced[14].

The solubility of the polymer bound peptide is of utmost importance in the application of the liquid-phase method to repetitive sequential type synthesis. The changes in the solubility and viscosity of poly(ethylene glycol)s when bound to various peptides are difficult to predict[26]. Thus, the solubility of the PEG esters of hydrophobic homooligomers $(Val)_n$ and $(Ile)_n$ decreases considerably for chain lengths $n > 6$. Conformational investigations reveal that a major reason for the exceptionally low solubility of these homooligopeptides is their tendency to form aggregated ß-structures with intermolecular hydrogen bonds[26, 27]. In the case of a number of other poly(ethylene glycol)-bound model peptides, this intimate relationship between physical properties and the conformation of the peptide chain could be clearly visualized[26, 28, 29].

A variation of the molecular weight of the support in the range of 2000–20000 shows no significant effect on the solubility of the attached peptides. Polyethers with molecular weights higher than 20000 have very low capacities whereas the lower polyethers are no longer amenable to crystallization. Most notably, low molecular weight polyethers can be used as solubilizing groups for the side-chain protection of peptides[30].

The synthesis of a number of model peptides and biologically active peptides clearly illustrates the feasibility of the liquid-phase method of peptide synthesis[31]. The use of anchoring groups between poly(ethylene glycol) and the peptide considerably facilitates the cleavage of the peptide from the support[32]. A photosensitive anchoring group has been used by Tjoeng et al[33]. These photosensitive poly(4-bromomethyl-3-nitrobenzoylethylene glycol) supports permit a mild photolytic release of the fully protected peptide under neutral conditions[33–35]. Poly(4-amino-methyl-3-nitrobenzoylethylene glycols) have been developed recently as soluble supports which permit photolytic release of peptide amides in good yields[36]. The attachment of these anchoring groups to the poly(ethylene glycol) backbone could be considerably facilitated using PEG with terminal amino groups[37, 38].

A significant variation of the liquid-phase method is the so-called liquid-solid-phase technique which combines the strategic features of the liquid- and solid-phase synthesis and the polymer reagent technique[39, 40]. The unique feature of this technique is the possibility of removing unreacted components after coupling by a simple filtration procedure. A combination between the liquid-phase strategy and the polymer reagent technique for stepwise peptide synthesis was proposed by Jung et al.[41, 42].

2.3.2 Nucleotide Synthesis

The repetitive sequential-type synthesis on soluble polymeric supports has also been applied to the synthesis of oligonucleotides. As the coupling kinetics in liquid-phase synthesis is identical to that of low molecular weight reactions, the yields of the condensation step resulting in the formation of phosphate ester bonds turned out to be high enough for the effective use of soluble polymers[43–45]. Linear polystyrenes have been used earlier as soluble polymeric supports for nucleotide synthesis[46, 47]. However, the considerable decrease in solubility during the synthesis turned out to be a serious drawback of this procedure; subsequently, soluble polymers with more favourable physiochemical properties (e. g. PEG) and with a high capacity (e. g. poly(vinyl alcohol) and poly(ethylenimine)) were used[43, 48]. The choice of the polymeric protecting group depends on the strategy for building up the nucleotides[49, 50]. For the stepwise synthesis, polymers with high capacity are preferred. In the case of the condensation of smaller fully protected segments to larger oligonucleotide chains, polymers with a low capacity such as PEG have proved to be most effective. Hydrolysis products of linear non-cross-linked copolymers from vinyl acetate and N-vinylpyrrolidone have also been used as supports for oligonucleotide syntheses in homogeneous solution[51].

2.4. Soluble Polymeric Reagents

The concept of soluble polymeric reagents permits to work in a homogeneous phase and should overcome the drawbacks characteristic of reactions in heterogeneous matrices[52–56]. The advantages of working in a homogeneous medium and thereby realizing maximum reaction yields are exploited in the design of a few soluble polymeric reagents for synthetic purpose and also in the preparation of soluble polymeric catalysts. The working up of the reaction mixture can be effected by selective precipitation or by ultrafiltration. Most notably, as the ultrafiltration procedure is becoming increasingly convenient[57, 58], the use of linear soluble polymers for reagents in organic synthesis is of growing interest.

2.4.1 Reagents Based on Polystyrene and Analogous Linear Polymers

In contrast to the use of insoluble cross-linked polystyrene as a carrier for organic reagents, there are very few reported examples of the use of linear polystyrene as a soluble carrier of organic reagents. Hallensleben[59] investigated the application of linear and cross-linked poly(p-vinylbenzoyl chlorides) as reagents to the conversion of aliphatic and aromatic carboxylic acids to their acid chlorides. These polymeric reagents were prepared starting from polystyrene of molecular weight 267000. In all these polymer analogous reactions, the yields of the incorporation of the reagents by various methods are invariably much higher in the case of the linear polymers than for the cross-linked polymers.

It can be seen from Table 4 that the conversion yields are much higher when linear instead of cross-linked polystyrene is used as carrier.

Poly [p-(ω-lithiumalkyl)styrenes] with n = 1 to 4 were synthesized by the reaction of the corresponding halogen compounds with n-butyllithium. The application of these organometallic polymers as metallating agents to halogen-containing

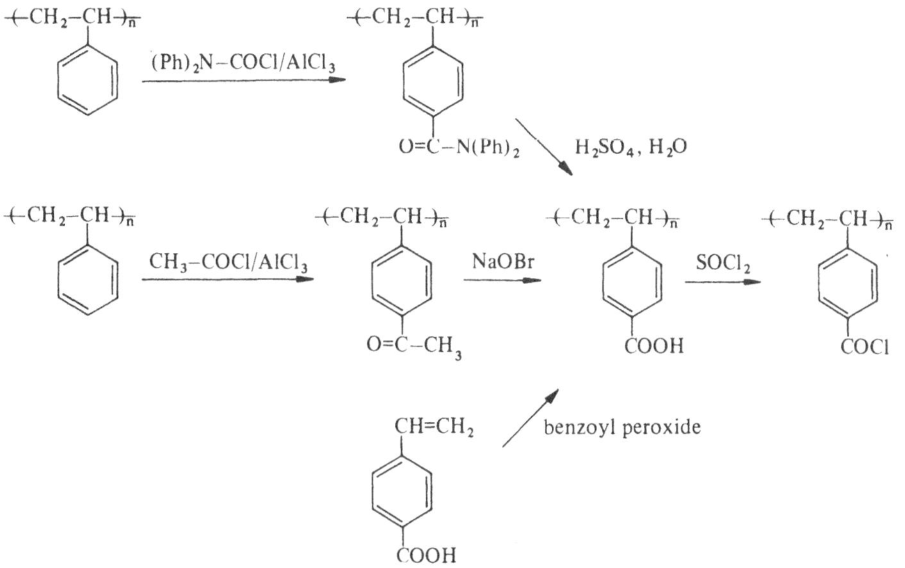

Scheme 3. Preparation of poly(p-vinylbenzoyl chloride). Poly(p-vinylbenzoyl chloride) prepared as indicated here converts acids into their chlorides in high boiling solvents (Table 4). The resultant acid chlorides can be separated from the reaction mixture by fractional distillation[59].

Table 4. Conversion of acids to acid chlorides

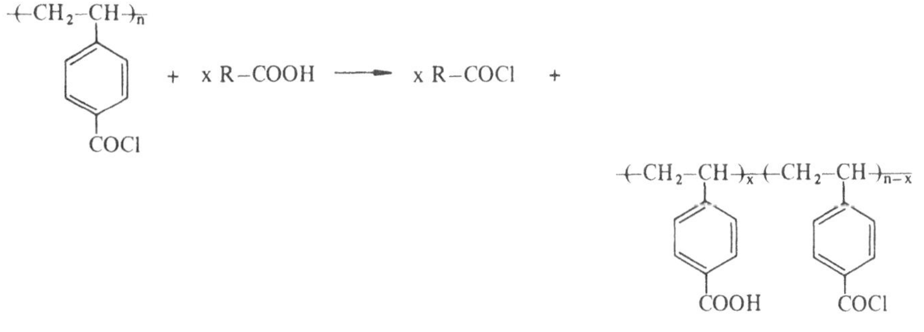

Substrate	Polymer reagent	Yield of acid chloride
Acetic acid	Linear	88
Benzoic acid	Linear	81
Caproic acid	Linear	61
Acrylic acid	Linear	57
Acetic acid	Cross-linked	50
Benzoic acid	Cross-linked	42

compounds and CH acidic substances has been investigated[60]. The metallation reactions utilizing these soluble polymers proceed in high yields even though the experimental techniques, after separation of the polymer, are somewhat difficult. The polymer is separated in these cases by selective precipitation with ether from a tetrahydrofuran or benzene solution.

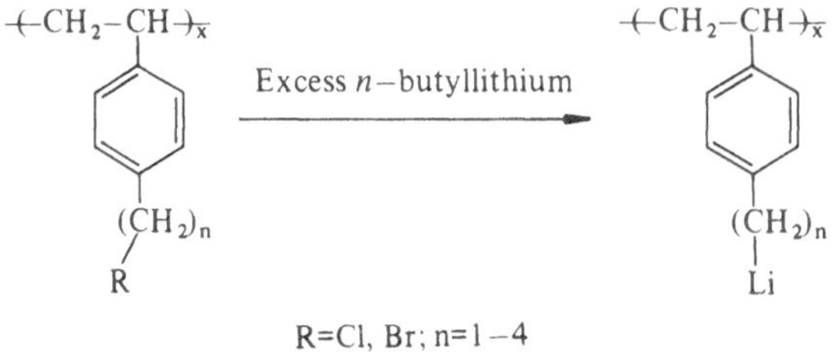

R=Cl, Br; n=1−4

Poly(4-vinylpyridine) boran has been used as a reducing agent for carbonyl compounds[61]. This polymeric reducing agent was prepared by the reaction of poly(4-vinylpyridinium hydrochloride) with sodium borohydride.

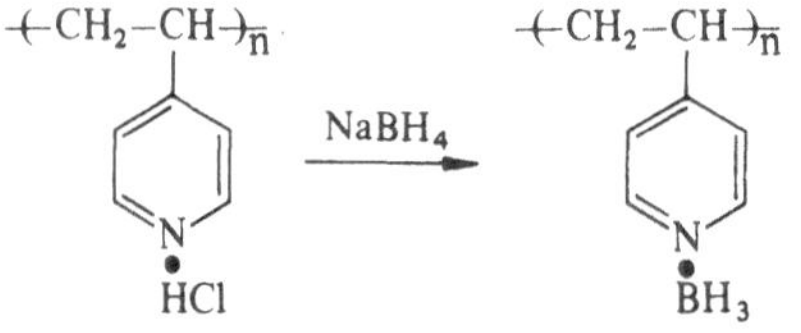

Linear polystyrene-bound tertiary phosphines have been alkylated and deprontonated to generate polymeric Witting reagents[62].

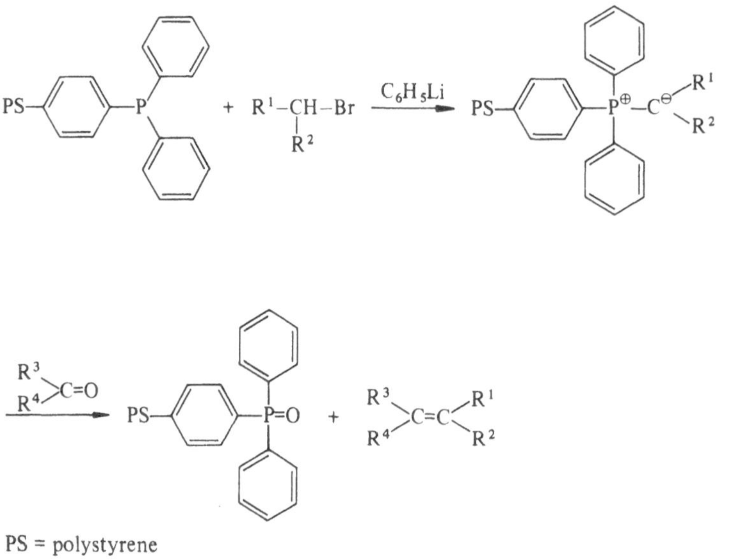

PS = polystyrene

Substituted olefins were obtained in equal yields compared to those obtained by analogous low molecular weight reactions. In contrast to Witting reagents on insoluble polymeric supports, these soluble polymeric Wittig reagents reveal the same kinetic behaviour as that of the low molecular weight system.

Linear soluble polystyrene-supported triphenylphosphine/carbon tetrachloride has been used as a condensing agent for the peptide synthesis in homogeneous solution [63, 64]. After the desired condensation, the polymer is precipitated quantitatively and removed by filtration. The efficiency of this technique has been demonstrated by the preparation of several dipeptide derivatives in 84–95% yields.

$$\text{(P)}-\langle\text{C}_6\text{H}_4\rangle-\text{P}(\text{C}_6\text{H}_5)_2 + \text{CCl}_4 + \text{Z}-\text{NH}-\overset{\text{R}'}{\underset{}{\text{CH}}}-\text{COOH} + \text{H}_2\text{N}-\overset{\text{R}''}{\underset{\text{H}}{\text{C}}}-\text{COOR}''' + 2\,\text{Et}_3\text{N}$$

$$\longrightarrow \text{(P)}-\langle\text{C}_6\text{H}_4\rangle-\overset{\text{O}}{\overset{\|}{\text{P}}}-(\text{C}_6\text{H}_5)_2 + \text{CHCl}_3 + \text{Z}-\text{NH}-\overset{\text{R}'}{\underset{}{\text{CH}}}-\text{CO}-\text{NH}-\overset{\text{R}''}{\underset{}{\text{CH}}}-\text{COOR}'''$$

Z = benzyloxycarbonyl

2.4.2 Reagents Attached to Polyamides

Chlorinated poly(hexamethyleneadipamides) (N-chloronylons) are readily soluble in a variety of solvents and are found to be effective oxidizing agents and chlorinating agents [65–68]. These N-chloronylons are prepared in good yields by chlorination with aqueous hypochlorous acid, t-butyl hypochlorite or chlorine monoxide.

$$-\overset{}{\underset{\text{H}}{\text{N}}}-\overset{}{\underset{\text{O}}{\text{C}}}-\text{R} \longrightarrow -\overset{}{\underset{\text{Cl}}{\text{N}}}-\overset{}{\underset{\text{O}}{\text{C}}}-\text{R}$$

Secondary alcohols, are oxidized to ketones in yields ranging from 62% to 97% and sulfides are oxidized to sulfones in 65–78% yields [65, 69]. Unsubstituted nylons regenerated during these reactions are insoluble in most solvents and precipitate in the reaction thus avoiding the separation problems normally associated with linear soluble polymers.

Oxidation of tertiary amines with N-chloronylon 6,6 has also been investigated in several solvents [70]. Thus, N-chloronylon has been found to oxidize N,N-dimethylaniline at room temperature to N-methylaniline in 15–50% yield, depending on the solvent used. The oxidation of N,N-dimethylbenzylamine gives benzaldehyde in 15–30% yield and a small amount of N-methylbenzylamine. A comparison of these oxidations with those using the low molecular weight analoges, e. g. N-chlorosuccinimide, indicates that in both cases the reaction paths are similar. However, the poor yields observed in these oxidations of amines make these polymeric oxidizing agents unsuitable for amine oxidations.

$$\left[\begin{matrix} \underset{O}{\overset{\parallel}{C}}-(CH_2)_4-\underset{O}{\overset{\parallel}{C}}-NH-(CH_2)_6-NH \end{matrix}\right]_n + 2\,(CX_3CO)_2O \longrightarrow$$

$$\left[\begin{matrix} \underset{O}{\overset{\parallel}{C}}-(CH_2)_4-\underset{O}{\overset{\parallel}{C}}-\underset{\underset{CX_3}{\overset{\mid}{CO}}}{\overset{\mid}{N}}-(CH_2)_6-\underset{\underset{CX_3}{\overset{\mid}{CO}}}{\overset{\mid}{N}} \end{matrix}\right]_n + 2\,CX_3COOH$$

$$X = Cl \text{ or } F$$

N,N′-Bis-(trichloroacetyl)-nylon 6,6 and N,N-bis-(trifluoroacetyl)-nylon 6,6 are prepared by the reaction of the corresponding substituted acetic anhydride with nylon 6,6[71, 72]. These trihalogenoacetyl polyamides are readily soluble in chlorinated hydrocarbon solvents, dioxane, tetrahydrofuran and DMSO.

N,N′-Bis(trifluoroacetyl)-nylon 6,6 has been used for the trifluoroacetylation of amines and alcohols[71, 73]. The reactions have been carried out in acetonitrile. In the course of the reaction, nylon 6,6 precipitates and can be separated by filtration. The corresponding trifluoroacetylamides are isolated in 41−96% yields. Trifluoroacetylation of alcohols preceeds in good yields[73]. The selectivity of this trifluoroacetylating agents to various substituted anilines and against structurally isomeric secondary amines has been studied in detail by kinetic measurements.

2.4.3 Poly(Ethylene Glycol)-Bound Reagents for Peptide Synthesis

Reagents attached to poly(ethylene glycol)s have been found to be useful in peptide synthesis[38, 74]. The terminal hydroxy groups of poly(ethylene glycol) are first converted to the more reactive primary amino groups and these groups are used for the attachment of the reagents to the polymer chain. Thus, PEG-bound active esters have been prepared in high yields by coupling successively 4-hydroxy-3-nitrobenzoic acid and carboxylic acid anhydride (or N-protected amino acid anhydride) to PEG−NH$_2$.

$$PEG-NH_2 + HO-\underset{}{\overset{NO_2}{\bigcirc}}-\overset{O}{\overset{\parallel}{C}}-OH \longrightarrow HO-\underset{}{\overset{NO_2}{\bigcirc}}-\overset{O}{\overset{\parallel}{C}}-NH-PEG \xrightarrow{(RCO)_2O}$$

$$R-\overset{O}{\overset{\parallel}{C}}-O-\underset{}{\overset{NO_2}{\bigcirc}}-\overset{O}{\overset{\parallel}{C}}-NH-PEG$$

$$PEG-NH_2 = H_2N-(CH_2-CH_2-O-)_nCH_2-CH_2-NH_2$$

These active PEG esters of the corresponding amino acids are readily soluble in CH$_2$Cl$_2$, DMF, pyridine, or water. The yields for the preparation of these reagents are considerably higher than those for the corresponding insoluble polymer reagents[75]. 1-Hydroxybenzotriazol (HOBt) esters have been prepared by the reaction of HOBt-5-carboxylic acid with PEG-NH$_2$ and subsequent treatment with the anhydride.

Poly(ethylene glycol)-bound carbodiimides have been prepared by the reaction of PEG-NH$_2$ with isocyanates[74]. These soluble polymeric carbodiimides have been used as dehydrating agents in peptide coupling and for the oxidation of alcohols to aldehydes and ketones. The application of these polymer-bound reagents combines the advantage of using a large excess of reagent in homogeneous solution with the separation of the polymeric component from the reaction product by selective precipitation.

2.4.4 Soluble Polymer-Bound Catalysts

The highly desirable combination of the homogeneity and the facility of the separation of the catalyst has been achieved in the design of soluble polymers attached to homogeneous transition metal cataysts[76–78]. In contrast to heterogeneous catalysis, homogeneous catalysis often exhibits higher substrate selectivity, better reproducibility and reactivity under mild conditions. The frequently encountered problem of separating low molecular weight catalysts from reaction products can be circumvented by binding the catalysts to soluble polymers. Thus, non-crosslinked linear polystyrene was chloromethylated and converted into macromolecular phosphine ligands by treatment with potassium diphenylphosphide[76, 79]. Equilibration of these soluble polymeric ligands with appropriate transition metal derivatives resulted in the formation of soluble macromolecular metal complexes, which catalyze the hydrogenation or hydroformylation of alkenes under mild conditions.

Continuous homogeneous catalysis is achieved by membrane filtration, which separates the polymeric catalyst from low molecular weight solvent and products. Hydrogenation of 1-pentene with the soluble polymer-attached Wilkinson catalyst affords n-pentane in quantitative yield[76, 78]. A variety of other catalysts have been attached to functionalized polystyrenes[76]. Besides linear polystyrenes, poly(ethylene glycol)s, polyvinylpyrrolidinones and poly(vinyl chloride)s have been used for the liquid-phase catalysis. Instead of membrane filtration for separating the polymer-bound catalyst, selective precipitation has been found to be very effective. In all cases, the catalysts can be recovered quantitatively and are recycled. The functionalization of poly(ethylene glycol) with the terminal metal centers of the Ziegler-Natta catalysts results in soluble polymeric catalysts which induce stereoregulated polymerization.

The soluble polymer-supported catalysts have also been used for asymmetrically catalyzed reactions[78]. Following a procedure for the preparation of insoluble polymeric chiral catalysts[80], a soluble linear polystyrene-supported chiral rhodium catalyst has been prepared. This catalyst displays high enantiomeric selectivity compared to the low molecular weight catalyst. Thus, hydroformylation of styrene using this catalyst produces aldehydes in high yields. The branched chiral hydrotropaldehyde is formed in 95% selectivity.

3 Soluble Polymeric Complexing Agents

In recent years, there has been much interest in the study of polymer complexes, particularly in relation to their catalyst properties, thermostability and biomedical effects. Mainly these studies have been directed towards the use of insoluble polymers. An account of the soluble polymeric metal complexes and their catalytic activity has been published recently[81]. A brief outline of the studies in relation to the metal-chelation of some linear hydrophilic polymers and their functional derivatives with chelating groups and some polymeric molecular complexes is given in this section.

3.1 Metal Chelation of Soluble Polymers

Soluble polymers such as poly(ethylenimine), polyvinylamine, poly(vinylsulfonic acid), poly(acrylic acid), and copolymers of 1-vinylpyrrolidone with other functional monomers have been functionalized with various well-known chelating groups[82-84]. The most important chelating groups employed are thiourea, 8-hydroxyquinoline and iminodiacetic acid. The resulting soluble chelating polymers have been found to be very selective in their chelation with metal ions. The binding capacity of these polymers are also high. Thus, 1 g of the polymer derived from polyethylenimine and thiourea binds 1 g of mercury. The application of these metal binding polymers for industrial waste water treatment can be foreseen[85].

Polyvinylamine and its derivatives have been also used for the complexation of metal ions in homogeneous phase[83]. Introduction of various chelating groups into polyvinylamine has been shown to result in a greater selectivity for complexation than polyvinylamine itself. Thus, poly(vinylaminodiacetate) has been found to be a selective complexing agent for copper and poly(N-methyl-N-vinylthiourea) for mercury[83].

Table 5. Some examples of soluble polymeric complexing agents

Structure unit of polymer ligand	Name of polymer	Metal ion	pH	Capacity (mg/g)	Molar capacity (mmole/g)
$-CH_2-CH_2-NH-$	Poly(ethylenimine)	Co^{2+}	4	105	1.8
		Ni^{2+}	4	135	2.3
		Cu^{2+}	4	180	2.8
		Cu^{2+}	4	185	1.6
$-CH_2-CH_2-N-$ $(CH_2)_2$ N CH_2 / CH_2 $COOH$ $COOH$	Poly(ethylenimine acetic acid)	Cu^{2+}	4	130	2.0
		Pd^{2+}	2.5	80	0.8
		Ag^+	2.5	40	0.4
$-CH_2-CH_2-N-$ $(CH_2)_2$ NH $S=C$ $NH-CH_3$	Poly(ethylenimine N-methyl N-thiourea)	Au^{3+}	2.5	180	0.9
		Pt^{4+}	2.5	135	0.7
		Hg^{2+}	4	100	0.5
$-CH_2-CH-$ .... $-CH_2-CH-$ N $=O$ CH_2 NH $S=C$ NH_2	Poly(1-vinyl-2-pyrrolidinone-co-allylthiourea)	Au^{3+}	2.5	172	0.9
		Hg^{2+}	4	92	0.5

Copolymerisation of N-vinylpyrrolidone or acrylic acid with other functional monomers such as allylic compounds and vinyl ethers results in the formation of polymers with excellent solubility properties and metal-ion binding capacities[82, 86].

$$\begin{array}{c}
\overset{}{\underset{\overset{|}{COOH}}{-(CH-CH_2)_m}} \overset{}{\underset{\overset{|}{CH_2}}{-(CH-CH_2)_n}} \\
\underset{\overset{|}{NH}}{} \\
\underset{\overset{\diagdown}{NH_2}}{S=C}
\end{array}$$

3.2 Charge Transfer Complexes with Iodine

Many of the synthetic soluble polymers form highly coloured charge transfer complexes with iodine. Such a charge transfer complex formation between metal polymer chelates and iodine is made use of in the design of some semiconductors. Although most of the semiconducting organic polymers so far developed (polyacetylenes, poly(4-vinylpyridine)iodine complexes, polyphthalocyanine-metal chelates) have good electrical properties, their practical use is limited because of the difficulties is processing them into electronic materials[87]. In order to prepare easily processible semiconducting polymers, Highashi and coworkers have attempted to use polymers which can be easily processed and chemically modified so as to obtain good electric conductivity[88]. Thus, they have succeeded in preparing new semiconductors from Cu^{++}-chelates of poly(vinyl alcohol) and polyacrylamide by charge transfer complexation with iodine[89, 90].

The poly(vinyl alcohol)-iodine charge transfer complexes have been the subject of many investigations[91-95]. Water-soluble iodine complexes with poly(1-vinylpyrrolidone) are commercially available as disinfecting agents[96]. Poly(N-vinyl carbazole)[97], polyamides[98] and poly(ethylene oxide) are also found to form molecular addition complexes with iodine and iodine compounds. An account of the different types of molecular complexes formed with various polymers has been published[99].

4 Pharmacologically Active Polymers

Various reasons have been put forward for the recent interest in the field of the pharmacologically active polymers. The main advantages of macromolecular drugs over simple compounds are their delayed action, sustained release, lower toxicity, choice of delivery, and potentiation of activity. Several reviews have been appeared describing the various applications of macromolecules in medicine[100-105].

Principally, three approaches can be visualized in the design of pharmacologically active polymers. One can take a preformed polymer and prepare the polymer drug either by complexing or covalently binding the drug to the polymer. A second approach is to prepare a polymerizable monomeric drug and then homopolymerize or copolymerize it with other monomers. A third approach involves the introduction

of the drug in a polymer matrix and allowing the drug to diffuse out into the biological fluids. The rate of drug diffusion desired can be controlled by the proper
choice of the polymers and the nature of the delivery system. Many of the structural
variables that affect the properties of polymers can certainly be expected to affect
the biological activity of a drug when the two are chemically associated. Thus, the
molecular weight, dimensional structures and chemical characteristics of the polymer
have definite effects on the pharmacological action. Among the chemical characteristics of the basis polymer intended as a drug, the solubility is of utmost importance.
The recent applications of soluble synthetic macromolecules as such as drugs and as
carriers of pharmacologically active compounds are described in this chapter.

4.1 Polymeric Quarternary Ammonium Antimicrobial Agents

The potentialities of quarternary ammonium germicides have been recognized very
early. The antibacterial action of this class of compounds is related to the interaction
of the cationic material with the cell membrane, with the release of enzymes and
other metabolic intermediates from the cell, leading to the destruction of the micro-
organism. A polymeric species with plurality of these cationic species could result
in a more potent antimicrobial drug, because of the availability of more points of
interaction with the cell wall. Bearing this in mind, Panarin and coworkers synthesized
various copolymers of N-vinylpyrrolidone and 2-methacryloyl-oxyethyl-N,N,N-triethyl-
ammonium bromide and iodide[106]. They found that the activity increased with
rising content of the quarternary ammonium moiety.

Cationic polymers produced by the reaction of polyepihalohydrins and various
amines have been found to be active as antimicrobial agents[107]. Cumarone-indene
polymers containing quarternary ammonium groups directly attached to the aryl
nucleii of the polymer possess antibacterial and fungicidal properties[108]. A number
of quarternary monomers, homopolymers and copolymers with antibacterial activity
have been reported by Samour and Richards[109]. The bacteriostatic and bactericidal
properties of homo- and copolymers of diethylammonium compounds have been

investigated[110]. The bactericidal actions of polyionenes, containing quarternary
ammonium groups located along the polymer chain, have been investigated[111].

4.2 Surface Active Polymers with Pharmacological Activity

Surface active poly(ethylene glycol) ethers have been found to exert a suppressive
effect on tuberculosis[112, 113]. The greatest activity is observed when the number of
ethylene oxide units is 15 to 20 and the activity is completely abolished in those
polymers where the number of units ranges between 25 and 30. Conforth and co-
workers have also found that when the number of ethylene oxide units was 45–70
the polymers, instead of displaying chemotherapeutic activity, significantly enhanced
the infection, the so-called protuberculous effect. Thus, as the lipophilic to hydrophilic ratio decreases, activity passes from antituberculosis to inactive protuberculous. These observations demonstrate the importance of two physical parameters
in chemotherapy, namely surface activity and molecular size.

The immunological properties of bovine serum albumin have been found to be
altered by covalent attachment to surface active poly(ethylene glycol)s. Thus,
Abuchowski and coworkers[114] attached monomethoxy poly(ethylene glycol)s of
molecular weight 1900 and 5000 g · mol^{-1} to bovine serum albumin using cyanuric
chloride. The resulting polymer-bound protein has been found to have lost its im-
munogenicity. The modified protein shows substantial changes in hydrodynamic
properties. The altered sedimentation constants and chromatographic properties
are consistent with the picture of a protein molecule surrounded by a flexible hydrophilic shell composed of poly(ethylene glycol) and its bound water. Such a shell
would cover antigenic determinants and render the albumin inert to immune processes. The observation that each ethylene oxide unit of poly(ethylene glycol) binds
approximately 3 molecules of water indicates that an albumin molecule with substantial amounts of poly(ethylene glycol) attached, possesses profoundly altered
hydrodynamic properties[115, 116].

The trisaccharide chain in Cinerubin A, an anthracycline antibiotic with anti-
mitotic activity, has been replaced by a poly(ethylene glycol)[117]. The resulting
poly(ethylene oxide)-bound ε-pyrromycinone is readily soluble in water and exhibits
full biological activity. In this case, the hydrophilic poly(ethylene oxide) chain
appears to be perfectly suitable to simulate the physicochemical properties of the
oligosaccharide chain.

The local anaesthetic action of procain, bound to tetraethylene glycol and PEG
400, has been investigated by Weiner and Zilkha[118]. They found the tetraethylene
glycol derivative displaying an extended and stronger activity.

4.3 Polymeric and Polymer-Bound Antibiotics

Smith and Marshall isolated and characterized polymeric materials formed in aqueous
solutions of a number of penicillin derivatives[119]. The degree of polimerization ranged

between 3.5 and 6.1. Polymers of α-aminobenzylpenicillin with molecular weights 1000–5000 have been prepared and tested for their relative activity; those with molecular weights in the range of 1000 to 3000 appeared to be most active[120].

Ushakov and Panarin[121, 122] synthesized polymeric salts of vinyl amine and vinyl alcohol copolymers and amides and hydrazides of penicillin derivatives. The coupled water-soluble, stable polymeric penicillin derivatives were found to have the same activity as the parent drugs.

The potentiation of tetracyclines relative to complexation with polyacrylic acid has been investigated by Takesue et al.[123]. In this case, the polymer has been found to promote greater absorption of the antibiotic into the bloodstream and exert a more powerful adjuvant action than simple acids such as citric acid.

An attempt to reduce the toxicity of streptomycin by attaching it to soluble polymers has been reported recently[124]. The formyl group of streptomycin has been condensed with methacrylic acid hydrazide in order to form a polymerizable derivative of streptomycin. The resulting hydrazone has been copolymerized with methacrylamide and 2-methylsulfinylethyl methacrylate to form water-soluble polymers. These copolymers show an increasing tuberculostatic activity.

4.4 Polymeric Anti-Cancer Agents

Water-soluble polymeric carriers based on the copolymerization of divinyl ether and maleic anhydride for the antitumor drug cyclophosphamide have been reported recently by Ringsdorf and coworkers[125, 126]. A cooperative action of the immunosuppressive nature of the alkylating antitumour drug cyclophosphamide and the immunostimulating nature of the polymeric carriers is presumed. Cyclophosphamide derivatives were also fixed on poly(4-glutamic acid) and linear poly(ethylenimine). The activities of cyclophosphamide and also of testosterone copolymerized with acrylamide, 2-methacryloyl-oxyethyltrimethylammonium chloride and N-vinyl-pyrrolidinone have also been investigated[127]. A spacer effect has been observed in the investigation of the activity of testosterone. Testosterone derivatives directly fixed to the polymer chain are inactive whereas those fixed via a diester as spacer are as active as the low-molecular weight analogues.

Polycationic species derived from poly(ethylenimine), poly(propylenimine), and polyvinylamine are shown to have tumour growth inhibiting properties[128, 129]. Selective anti-tumour therapy with various amido-amine polymers was investigated by Ferruti et al.[130, 131].

4.5 Polymeric Radioprotecting Agents

The higher radiation protectiveness of polymeric compounds over low molecular weight compounds could be utilized in the design of radioprotective agents. A report dealing with the various polymeric antiradiation substances has been published[132]. A number of water-soluble polymeric radioprotective agents have been prepared by

$$+CH_2-CH)_{\overline{n}} ----+CH_2-CH)_{\overline{m}}$$

with NH, C=O, O, X substituents on first unit; pyrrolidone (N–C=O ring) on second unit.

$$+CH_2-CR)_{\overline{n}} -----+CH_2-CR)_{\overline{m}}$$

first unit: C=O, O, X; second unit: C=O, $(CH_2)_2$, $N^{\oplus}(CH_3)_3Cl^{\ominus}$

$$+CH_2-CR)_{\overline{n}} ---- +CH_2-CR)_{\overline{m}}$$

first unit: C=O, O, X; second unit: C=O, NH_2

X = Cyclophosphamide or testosterone derivative

copolymerization of monomers containing potential radioprotective residues and
1-vinyl-2-pyrrolidone[133–135]. Water-soluble radioprotective polymers have also been
prepared by copolymerization of a 3-thiazolidone derivative and 1-vinylpyrrolidone[136].

$$+CH-CH_2)_{\overline{n}} --+CH_2-CH)_{\overline{m}}$$

first unit: C=O, O, $(CH_2)_2$, N, thiazolidone ring (O=, S, phenyl); second unit: pyrrolidone (N–C=O ring)

4.6 Miscellaneous

Soluble polymers have also been investigated as synthetic substitutes of plasma.
Besides poly(1-vinylpyrrolidone) which has been used very early for this purpose[136, 137], other synthetic polymers such as poly(vinyl alcohol) and
poly [N-(2-hydroxypropyl) methacrylamide]

$$\left[-CH_2-\underset{\underset{O=C-NH-CH_2-CH(OH)-CH_3}{|}}{\overset{\overset{CH_3}{|}}{C}}- \right]_n$$

have been proposed as substitutes of plasma. Copolymers from ethylene oxide and propylene oxide with a molecular weight between 50000–100000 have also been suggested as plasma substitutes[138].

Kropachev and coworkers have prepared copolymers of N-methacryloyl-4-aminobenzenesulfonamide with 1-vinylpyrrolidone and shown that these copolymers display prolonged antibacterial activity[139].

A synthesis of prototypes for natural retinal-protein complexes has been attempted by Bayer and coworkers[140]. They attached retinal, which is the chromophoric group of the protein complexes rhodopsin and retinochrome, to a modified poly-(ethylene oxide) and investigated the red shift of these complexes. To imitate the natural binding of the retinal, oligopeptides with identical functional side chains were incorporated into the system. The red shift behaviour of this semisynthetic polymeric product was found to be the same as that of the natural product.

5 Immobilized Enzymes and Synzymes

Polymers have been used since the last two decades for the immobilization of enzymes, and a number of articles have been published on the application of insoluble polymeric supports for the immobilization of enzymes[5, 141–143]. Besides this application of polymers in enzyme technology, very recently there has been much enthusiasm in the development of synthetic polymers with catalytic activity mimicing that of enzymes. Such synthetic polymers are now known as synzymes.

5.1 Immobilization of Enzymes on Soluble Polymers

Most of the earlier studies on the immobilization of enzymes were directed towards the attachment of the enzymes to water-insoluble polymeric supports such as cellulose[144], dextran derivatives[145], polyacrylamide[146] and porous glass[147].

Diffusion problems and steric hindrance are two main factors affecting the application of such supports. The introduction of soluble polymers for immobilization purposes overcomes these difficulties to a greater extent. These soluble enzyme derivatives were synthesized in order to increase the effective molecular size of parent enzymes; this would permit the use of ultrafiltration without any loss of the enzyme. O'Neill et al. immobilized the enzyme chymotrypsin on soluble dextran for use in casein hydrolysis in an ultrafilter reactor[148]. Wykes and coworkers investigated the immobilization of α-amylase on soluble polymeric supports[149]. They synthesized the soluble amino-s-triazinyl derivatives of dextran and CM-cellulose by reaction with 2-amino-4,6-dichloro-s-triazine. These soluble immobilized amylase enzymes exhibited 67% of the specific activity of the free enzyme; the same enzyme immobilized on various insoluble supports showed very little activity (3–16%)[150–153]. 2-Ethyl-5-phenylisoxazolium-3-sulfonate (Woodward's reagent K) has been employed as the coupling agent for the fixation of amylase to soluble polymers such as poly-(acrylic acid) and poly(4-glutamic acid)[154, 155].

Water-soluble enzyme derivatives have been reported for poly(methacrylic acid anhydride)[156] and for poly(maleicanhydrideethylene)[157]. Monomethoxy poly-(ethylene glycol)s of molecular weights 1900 and 5000 g · mol^{-1} were covalently attached to bovine liver catalase using 2,4,6-trichloro-s-triazine as the coupling agent. The resulting PEG-1900-catalase retained 93% of its enzymic activity and PEG-5000 retained 95% of the activity. PEG-5000-catalase resisted digestion by trypsin, chymo-trypsin and a protease from *Streptomyces griseus*[158].

The activities of free and poly(1-vinylpyrrolidone)-bound trypsin were compared by Specht et al[159]. They observed a significantly lower degree of inactivation (self-digestion) of the polymer-bound enzyme than the unbound enzyme. Kallikrein has also been attached to poly(1-vinylpyrrolidone)[160].

Investigations on the spacer function and the cleavability of the substrate from water-soluble copolymeric enzymes based on poly[N-(2-hydroxypropyl)methacryl-amide] were carried out by Kopecek and coworkers[161, 162]. These studies reveal that there is no simple relationship between the length of the spacer and the cleava-bility of the substrate. It has also been observed that the length and the type of the spacer allows the kinetics of the cleavage to be controlled.

5.2. Synzymes

Synzymes are synthetic polymers with catalytic activity mimicing that of enzymes. Numerous investigations in search of synzymes have been reported[163–167]. The main approach for the design of synzymes has been to modify polymers and mold their conformation to markedly increase their affinity for small molecules.

Synthetic polyelectrolytes as models for enzymes have been reported recently[168]. The hydrolysis of dextrin in the presence of copolymers of vinyl alcohol and vinyl-sulfonic acid has been found to increase with increasing content of vinyl alcohol in the copolymers[169].

This acceleration has been explained by the hydrogen-bonding interaction of the copolymer with the substrate. Poly(p-styrenesulfonic acid-co-acrylic acid) has been found to catalyze the hydrolysis of amylose and sucrose[107]. Partially o-benzylsulfonated poly(vinyl alcohol) has been found to catalyze the hydrolysis of alkyl esters[171].

Hydrolysis of peptides and proteins has been observed to be catalyzed by poly(ethylenesulfonic acid) and poly(p-styrenesulfonic acid)[172, 173]. Poly(vinylbenzyltriethylammonium hydroxide) has been found to catalyze the alkaline hydrolysis of aliphatic esters[174].

References

1. For recent reviews on solid-phase synthesis, see: Erickson, B. W., Merrified, R. B.: Proteins, 3rd Ed. 2 (1976); Blossey, E. C., Neckers, D. C., (Editors): Benchmarks in chemistry — solid phase synthesis, Dowden, Hutchinson & Ross, Penn., 1975; Birr, Ch.: Aspects of the Merrifield Peptide Synthesis, Springer-Verlag, Heidelberg 1979
2. Neckers, D. C.: J. Chem. Educ., 52, 695 (1975)
3. Overberger, C. G., Sannes, K. N.: Angew. Chem. 86, 139 (1974); Leznof, C. C., Chem. Soc. Rev. 3, 65 (1974); Patchornik, A., Kraus, M. A., in: Encyclopedia of Polymer Science and Technology (Bikales, N. M. (Editor)), Wiley, New York, 1976. Suppl. 1, p. 468
4. Manecke, G., Storck, W.: Angew. Chem., 90, 691 (1978)
5. Goldman, R., Goldstein, L., Katchalski, E., in: Biochemical Aspects of Reactions on Solid Supports, (G. R. Stark, (Ed.)), Academic Press, New York, 1974; Goldstein, L., Manecke, G., in: Applied Biochemistry, (L. B. Wingard Jr., Katchalski-Katzir, E., Goldstein, L., (Editors)), Academic Press, New York, 1976, Vol. 1. p. 23
6. Batz, H. G.: Adv. Polym. Sci., Vol. 23 (H. J. Cantow et al., (Ed.)), Springer-Verlag, Heidelberg, Berlin (1977), p. 25
7. Leznof, C. C.: Acc. Chem. Res. 11, 327 (1978)
8. Crowley, J. I., Rapoport, H.: Acc. Chem. Res. 9, 135 (1976)
9. Heitz, W.: Adv. Polym. Sci., Vol. 23 (H. J. Cantow et al., (Ed.)), Springer Verlag, Berlin (1977), p. 1
10. Frechet, J. M. J., Farrall, M. J., in: Chemistry and Properties of Cross-linked Polymers (Labana, S. S. (Ed.)), Academic Press (1977)
11. Flory, P. J.: J. Am. Chem. Soc. 61, 3334 (1939)
12. Bayer, E. et al.: J. Am. Chem. Soc. 96, 7333 (1974)
13. Andreatta, R. H., Rink, H.: Helv. Chim. Acta 56, 1205 (1973)
14. Shemyakin, M. M. et al.: Tetrahedron Lett. 2323 (1965)
15. Green, B., Garson, L. R.: J. Chem. Soc. (C), 401 (1969)
16. Morawetz, H., in: Peptides: Chemistry, Structure and Biology (Walter, R., Meienhofer, J. (Eds.)), Ann Arbor Sci. Publ., Michigan (1975), p. 385–394
17. Blecher, H., Pfaender, P.: Liebigs Ann. Chem., 1263 (1973)
18. Geckeler, K., Bayer, E.: Liebigs Ann. Chem. 1671 (1975); Geckeler, K., Bayer, E.: Makromol. Chem. 175, 1995 (1974)
19. Bayer, E., Mutter, M.: Nature 237, 512 (1972)
20. Mutter, M., Hagenmeier, H., Bayer, E.: Angew. Chem. Int. Ed. Engl. 10, 811 (1972)
21. Mutter, M., Bayer, E.: Angew. Chem. Int. Ed. Engl. 13, 88 (1974)
22. Mutter, M. Uhmann, R., Bayer, E.: Liebigs Ann. Chem. 901 (1975)
23. Bayer, E. et al.: Tetrahedron 34, 1829 (1978)
24. Mutter, M.: Habilitationsschrift, Univers. Tübingen (1976)

25. Neidlinger, H., Dissertation, Univers. Mainz, Germany (1975)
26. Rahman, S. A., Anzinger, H., Mutter, M.: Biopolymers, (in press)
27. Toniolo, G., Bonoro, G. M., Palumbo, M., Pysh, E. S., in: Peptides 1976 (Loffet, A., Ed.)
    pp. 597–600, Univ. de Bruxelles, Brussels (1976)
28. Mutter, M.: Macromolecules *10*, 1413 (1977)
29. Imae, T., Ikeda, S.: Biopolymers *11*, 509 (1972)
30. Anzinger, H., Mutter, M., Bayer, E., Angew. Chem. *91*, 747 (1979)
31. Bayer, E., Mutter, M.: Chem. Ber. 107, 1344 (1974); Frank, H., Hagenmeier, H.: Tetrahe-
    dron *30*, 2523 (1974); Mutter, M. et al.: Biopolymers *15*, 917 (1976); Mutter, H., Mutter, M.,
    Bayer, E.: Z. Naturforsch., *34b*, 874 (1979); Goehring, W., Jung, G.: Liebigs Ann. Chem.
    1965 (1975); Weber, U., Hoppe-Seyler's Z. Physiol. Chem. *356*, 701 (1975); Frank, H.,
    Hagenmeier, H., Beckmann Rep. 12–17 (1974); Hagenmeier, H.: Hoppe-Seyler's Z. Physiol.
    Chem. *356*, 777 (1975); Stein, W., Dissertation Univers. Tübingen, Germany (1975)
32. Künzi, H.: Ger. Off. 2435642, Appl. P. 2435642. 1, 24 pp.
33. Tjoeng, F. S. et al.: Biochim. Biophys. Acta *490*, 489 (1977)
34. Tjoeng, F. S., Tong, E. K., Hodges, R. S.: J. Org. Chem. *43*, 4190 (1978)
35. Tjoeng, F. S. Hodges, R. S.: Tetrahedron Lett. 1273 (1979)
36. Pillai, V. N. R., Mutter, M., Bayer, E.: ibid. 3409 (1979)
37. Pillai, V. N. R.: Synthesis 23 (1980)
38. Mutter, M.: Tetrahedron Lett. 2839 (1978); Geckeler, K.: Polym. Bull. *1*, 427 (1979)
39. Frank, H., Hagenmaier, H.: Experientia *31*, 131 (1975)
40. Frank, H., Meyer, H., Hagenmaier, H., in: Peptides: Chemistry, Structure and Biology
    (Walter R., Meienhofer, J. (Eds.)), Ann. Arbor Sci. Publ. Michigan (1975), p. 439
41. Jung, G. et al., in: Peptides: Chemistry, Structure and Biology (Walter, R., Meienhofer, J.
    (Eds.)), Ann. Arbor Sci. Publ. Michigan (1975), p. 433
42. Heusel, G. et al.: Angew. Chem. Int. Ed. Engl. *16*, 142 (1977)
43. Schott, H., Brandstetter, F., Bayer, E.: Makromol. Chem. *173*, 247 (1973)
44. Brandstetter, F., Schott, H., Bayer, E.: Tetrahedron Lett. 2997 (1973)
45. Köster, H.: ibid. 1535 (1972)
46. Hayatsu, H., Khorana, H. G.: J. Am. Chem. Soc. *88*, 3182 (1966)
47. Cramer, F. et al.: Angew. Chem. *78*, 640 (1966)
48. Schott, H.: Angew. Chem. Int. Ed. Engl. *12*, 246 (1973)
49. Ohtsuka, E. et al.: J. Am. Chem. Soc. *92*, 3441 (1970)
50. Cook, A. F.: J. Am. Chem. Soc. *92*, 190 (1970)
51. Seliger, H., Aumann, G.: Makromol. Chem. *176*, 609 (1975)
52. Collman, J. P., Reed, C. A.: J. Am. Chem. Soc. *95*, 2048 (1973)
53. Beyerman, H. C., de Leer, E. W. B., van Vossen, W.: J. Chem. Soc., Chem. Common.,
    929 (1972)
54. Crowley, J. I., Harvey, T. B., Rapoport, H.: J. Makromol. Sci. Chem. *7*, 1118 (1973)
55. Crowley, J. I., Rapoport, H.: J. Amer. Chem. Soc. *92*, 6363 (1970)
56. Scott, L. T. et al.: J. Am. Chem. Soc. *99*, 625 (1977)
57. Bayer, E. Schurig, V.: Angew. Chem. *87*, 484 (1975)
58. Brandstetter, F., Schott, H., Bayer, E.: Makromol. Chem. *176*, 2163 (1975)
59. Hallensleben, M. L.: Angew. Makromol. Chem. *31*, 143 (1973)
60. Hallensleben, M. L.: ibid. *31*, 147 (1973)
61. Hallensleben, M. L.: Z. Naturforsch. *28b*, 540 (1973)
62. Kriech, W.: Dissertation, Univers. Tübingen, Germany (1977)
63. Appel, R., Willms, L.: J. Chem. Res. (S) 84 (1977); Appel, R., Willms, L.: ibid (M)
    901 (1977)
64. Appel, R., Wihler, H. D.: Chem. Ber. *109*, 3446 (1976)
65. Schuttenberg, H., Schulz, R. C.: Angew. Chem. Int. Ed. Engl. *10*, 856 (1971)
66. Schuttenberg, H. et al.: J. Makromol. Sci. Chem. *7*, 1085 (1973)
67. Hahn, K., Schulz, R. C.: Angew. Makromol. Chem., *50*, 53 (1976)
68. Yamaguchi, H., Schulz, R. C.: Makromol. Chem. *177*, 3441 (1976)
69. Sato, Y., Kunieda, N., Kinoshita, M.: Chem. Lett., 1023 (1972)

70. Sato, R., Schulz, R. C.: Makromol. Chem. *180*, 299 (1979)
71. Günster, E., Schulz, R. C.: Makromol. Chem. *179*, 2583 (1978)
72. Schuttenberg, H., Schulz, R. C.: Angew. Chem. *88*, 848 (1976)
73. Günster, E. J., Schulz, R. C.: Makromol. Chem. *180*, 1891 (1979)
74. Mutter, M.: Tetrahedron Lett. 2843 (1978)
75. Kalir, R. et al.: Eur. J. Biochem. *59*, 55 (1975)
76. Bayer, E., Schurig, V.: Angew. Chem. Int. Ed. Engl. *14*, 493 (1973)
77. Bayer, E., Schurig, V.: Ger. Pat. Appl. DOS *2*, 326, 489 (1973)
78. Bayer, E., Schurig, V.: Chem. Technol. *6*, 212 (1976)
79. Pittman, C. U. Jr., Evans, G. O.: ibid. *3*, 560 (1973)
80. Dumont, W. et al.: J. Am. Chem. Soc. *95*, 8295 (1973)
81. Tsuchida, E., Nishide, H.: Adv. Polym. Sci. *24*, 1 (1977)
82. Geckeler, K. et al.: Pure Appl. Chem. *52*, 1883 (1980)
83. Geckeler, K., Weingärtner, K., Bayer, E.: Makromol. Chem. *181*, 585 (1980); Bayer, E., Geckeler, K., Weingärtner, K., in: Polymeric Amines and Ammonium Salts, (E. Goethals, Ed.), Pergamon Press, Oxford 1980, in press
84. Welzmann, N.: J. Polym. Sci. *33*, 377 (1958)
85. Mutter, M., Bayer, E.: Chem. Technol., in press; Bayer, E., Weingärtner, K., Geckeler, K.: Makromol. Chem., in press
86. Geckeler, K., Mutter, M.: unpublished results
87. Katon, J. E.: Organic Semiconducting Polymers, Marcel Dekker, New York, 1976
88. Kakinoki, H. et al.: J. Polym. Sci. Polym. Lett. Ed. *14*, 407 (1976)
89. Higashi, F. et al.: ibid. *15*, 2303 (1977)
90. Higashi, F., Cho, C. S., Kakinoki, H.: J. Polym. Sci. Polym. Chem. Ed. *17*, 313 (1979)
91. Staudinger, H., Frey, K., Stark, W.: Ber. dtsch. chem. Ges. *60*, 1782 (1927)
92. Hermann, W. O., Hochnel, W.: ibid. *60*, 1658 (1927)
93. Kikukowa, K., Nozakura, S., Murahashi, N.: Polym. J. *3*, 52 (1972)
94. Inagaki, F. et al.: Bull. Chem. Soc. Japan *45*, 3384 (1972)
95. Heyde, M. et al.: J. Amer. Chem. Soc. *94*, 5222 (1972)
96. Commercial name: Polyvidon, Melsungen, Germany, PVP-Jod, Alsdorf/Germany
97. Hermann, A. M., Rembaum, A.: J. Polym. Sci. C *17*, 107 (1967)
98. Matsubara, I., Magill, J. H.: Polymer, *7*, 199 (1966)
99. Schulz, R. C.: Pure Appl. Chem. *38*, 227 (1974)
100. Lymann, D. J.: Rev. Macromol. Chem. *1*, 355 (1966)
101. Mark, H. F.: Pure Appl. Chem. *16*, 201 (1968)
102. Kovanov, V. V.: Usp. Khim. Technol. Polim. 176 (1970)
103. Merigan, T. C.: Nature *214*, 416 (1967)
104. Ferruti, P.: Pharmacol. Res. Comm. *7*, 1 (1975)
105. Ringsdorf, H., in: Polymeric Delivery Systems (Kostelnik, R. J. (Ed.)), Midland Macromolecular Monographs, Vol. 5, p. 197, Gordon & Beach, New York, 1978
106. Panarin, E. F., Solovski, M. V., Ekzenphylarov, O. N.: Khim. Pharm. Zh. *5*, 24 (1971); Chem. Abstr. *75*, 152274 d (1971)
107. Legator, M., C. A.: US Pat. 3,567,420 (1971)
108. Lane, E. W.: US Pat. 2,806,019 (1957)
109. Samour, C. M., Richards, M. C.: US Pat. 3,778,693 (1973); 3,879,447 (1975); 3,936,492 (1976)
110. Hoover, M. F.: US Pat. 3,539,684 (1970)
111. Rembaum, A.: Appl. Polym. Symp. *22*, 299 (1973)
112. Conforth, J. W., D'Arcy Hart, P., Stock, J. A.: Nature *168*, 150 (1951)
113. Conforth, J. W., Morgan, E. D., Potts, K. T.: Tetrahedron *29*, 1659 (1973)
114. Abuchowski, A. et al.: J. Biol. Chem. *252*, 3578 (1977)
115. Liu, K. J., Parsons, J. L.: Macromolecules *2*, 529 (1969)
116. Maxfield, J., Shepherd, I. W.: Polymer *16*, 505 (1975)
117. Geckeler, K., Mutter, M.: Z. Naturforsch. *34b*, 23 (1979)
118. Weiner, B. Z., Zilkha, A.: J. Med. Chem. *16*, 573 (1973)

119. Smith, H., Marshall, A. C.: Nature, *232*, 45 (1971)
120. Butcher, B. T. et al.: Mol. Cryst. Liqu. Cryst. *12*, 321 (1971)
121. Ushakov, B. N., Panarin, E. F.: Dokl. Akad. Nauk. USSR *149*, 334 (1963)
122. Ushakov, B. N., Panarin, E. F.: ibid. *147*, 5 (1962)
123. Takesue, E., Hlavka, J. J., Boothe, J. H.: US Pat., 3,356,571 (1967)
124. Hofmann, V., Ringsdorf, H.: Makromol. Chem. *180*, 595 (1979)
125. Hirano, T., Klesse, W., Ringsdorf, H.: Makromol. Chem. *180*, 1125 (1979)
126. Batz, H. G., Ringsdorf, H., Ritter, H.: Makromol. Chem. *175*, 2229 (1974)
127. Ringsdorf, H., Forschungsbericht, BMFT FBT *76*, 36-86-98 (1976)
128. Furhan, M., Moroson, H., in: Radiation, Protection and Sensitization (Moroson, H., Quintiliani (Eds.)), Taylor and Francis, p. 233 (1975)
129. Moroson, H., in: Chemotherapy of Cancer Dissemination and Metastasis (Garattini, S., Franchi, G., (Eds.)), Raven Press, New York p. 245, 1973
130. Ferruti, P. et al.: J. Med. Chem. *16*, 496 (1973)
131. Ferruti, P. et al.: J. Med. Chem., (in press)
132. Ringsdorf, H.: Strahlentherapie *132*, 627 (1967)
133. Barnes, J. H. et al.: Eur. J. Med. Chem. *12*, 467 (1977)
134. Overberger, C. G., Ringsdorf, H., Avchen, B.: J. Org. Chem. *30*, 3088
135. Overberger, C. G., Ringsdorf, H.: J. Med. Chem. *8*, 862 (1965)
136. Montoya, R. et al.: Makromol. Chem. *178*, 1221 (1977)
137. Heet, G., Wesse, H.: Münch. Med. Wochenschr. *85*, 11 (1943)
138. Wyandotte Chem. Corp., S. African Pat. 6,805,978; Chem. Abstr. *72*, 39088 (1970)
139. Shehukovskaya, L., Kapustayanskaya, A. M., Kropachev, V. A.: USSR Pat. 3,281,10 (1970); Chem. Abstr. *77*, 20349 (1972)
140. Das, P. K. et al.: J. Amer. Chem. Soc. *101*, 239 (1979)
141. Manecke, G.: Naturwissenschaften *51*, 25 (1964)
142. Zabovsky, O.: Immobilized enzymes, CRC Press, Cleveland, 1974
143. Silman, I. H., Katchalski, E.: Ann. Rev. Biochem. *35*, 873 (1966)
144. Smirnov, B. P., Manoilov, S. E.: Biokhimiya *31*, 387 (1966)
145. Axen, R., Panath, J., Ernbach, S.: Nature *214*, 1302 (1967)
146. Mosbach, K.: Acta. Chem. Scand. *24*, 2093 (1970)
147. Weetall, H.: Science *166*, 615 (1969)
148. O'Neill, S. P. et al.: Biotech. Bio Eng. *13*, 319 (1971)
149. Wykes, J. R., Dunnill, P., Lilly, M. D.: Biochim. Biophys. Acta *250*, 522 (1971)
150. Manecke, G.: Pure Appl. Chem. *4*, 507 (1962)
151. Barker, S. A., Somers, P. J., Eptom, R. E.: Carbohydr. Res. *8*, 491 (1968)
152. Barker, S. A., McLaren, J. V.: ibid. *14*, 287 (1970)
153. Ledingham, W. M., Hornby, W. E.: FEBS Letters *5*, 118 (1969)
154. Patel, R. P. et al.: Biopolymers *5*, 577 (1967)
155. Mitz, M. A., Sumania, L. J.: Nature *189*, 576 (1961)
156. Conte, A., Lehmann, K.: Hoppe-Seyler's Z. Physiol. Chem. *352*, 533 (1971)
157. Levin, Y. et al.: Biochemistry *3*, 1905 (1964)
158. Abuchowski, A. et al.: J. Biol. Chem. *252*, 3582 (1977)
159. Von Specht, B. U., Seinfeld, H., Brendel, W.: Hoppe-Seyler's Z. Physiol. Chem. *354*, 1659 (1973)
160. Von Specht, B. U. et al.: Arch. Inst. Pharm. *213*, 242 (1975)
161. Kopecek, J., Bozilova, H.: Eur. Polym. J. *9*, 7 (1973)
162. Drobnik, J. et al.: Makromol. Chem. *177*, 2833 (1976)
163. Katchalski, E. et al.: Arch. Biochim. Biophys. *88*, 361 (1960)
164. Letsinger, R. L., Savereide, T. J.: J. Am. Chem. Soc. *84*, 3122 (1962)
165. Sheehan, J. C., Bennet, G. B., Schneider, J. A.: J. Am. Chem. Soc. *88*, 3455 (1966)
166. Sakurada, I. et al.: Makromol. Chem. *91*, 243 (1966)
167. Morawetz, H. et al.: J. Am. Chem. Soc. *90*, 651 (1968)
168. Okubo, T., Ise, N.: Adv. Polym. Sci. *25*, 136 (1977)
169. Arai, K., Ise, N.: Makromol. Chem. *176*, 37 (1975)

170. Arai, K., Hagiwara, N., Ise, N.: Nippon Kagoku Kaishi, 201 (1975)
171. Suzuki, K. et al.: Prepr. 21st Polym. Symp. 1045 (1972)
172. Kern, W., Herold, W., Scherhag, B.: Makromol. Chem. *17*, 231 (1955)
173. Kern, W., Scherhag, B.: ibid. *28*, 209 (1958)
174. Arcus, C. L., Gonzalez, C. G., Linnecar, D. F. C.: Chem. Comm. 1377 (1969)

Received December 18, 1979 / April 18, 1980
H. J. Cantow (editor)

Flow and Electric Birefringence in Rigid-Chain Polymer Solutions

Victor N. Tsvetkov and Larisa N. Andreeva

Institute of Macromolecular Compounds, USSR Academy of Science, 199004 Leningrad, USSR

Results of the investigations of flow and electric birefringence in solutions of some rigid-chain polymers – cellulose derivatives, ladder polymers, aliphatic and aromatic polyamides etc. – are reported and discussed. In their dynamooptical and electrooptical properties, these polymers exhibit some specific features as compared with flexible-chain polymers. Flow birefringence and electric birefringence are efficient methods for studying the structural and conformational characteristics of the macromolecules of rigid-chain polymers.

Table of Contents

1 Introduction

1.1 Chain Flexibility

In recent years much attention in the field of polymer science and technology has been devoted to rigid-chain polymers. This is due to the fact that many currently used polymer materials with very valuable thermomechanical properties are based on macromolecular compounds characterized by limited chain flexibility; usually, these compounds are called rigid-chain polymers.

Speaking of polymer chain flexibility we mean the properties of an "isolated" single molecule exhibited in a dilute polymer solution. Under these conditions, two concepts of flexibility (and, hence, of rigidity) of chain molecules may be distinguished: equilibrium (static) flexibility and kinetic flexibility.

The equilibrium flexibility (or rigidity) of a chain characterizes an "average" conformation of the molecule being in the equilibrium state in a dilute solution. The information concerning this flexibility can be obtained by studying the size and shape of the macromolecule under these conditions. The kinetic flexibility (or rigidity) characterizes the rate of the transition of a chain molecule from one conformation to another. The time required for a change in the conformation of the molecule is a quantitative measure of its kinetic rigidity.

The problem of the relationship between the equilibrium and the kinetic rigidity of the chain is of paramount importance since both the behavior of a chain molecule in solution and the main properties of polymer materials are related to these molecular characteristics.

When a chain molecule is represented by an equivalent Gaussian chain of freely jointed segments[1, 2], the length of the Kuhn segment A may serve as a quantitative measure of the equilibrium rigidity of the chain. It is determined by the equation

$$\langle h^2 \rangle = AL \tag{1}$$

where L is the contour (hydrodynamic) chain length and $\langle h^2 \rangle$ the square and-to-end distance averaged over all conformations.

The "persistent" or "worm-like" Porod chain[3], a line in space the curvature of which is the same at all points and which depends on the "persistent length", a, is a more universal model for describing the conformational properties of a chain molecule. This length characterizes the distance along the chain at which the correlations in the orientations of successive chain units extend. It is determined by

$$\langle \cos \psi \rangle = \exp(-L/a) \tag{2}$$

where L is the contour length of a worm-like chain curved in such a manner that the angle between the chain directions at its beginning and at its end is ψ and $\langle \cos \psi \rangle$ is the value of $\cos \psi$ averaged over all chain conformations.

The mean square end-to-end distance for a worm-like chain, $\langle h^2 \rangle$, and its radius of gyration, $\langle R^2 \rangle$, are related to the contour length of the chain by Eqs. (3) and (3')[3, 4]:

$$\langle h^2 \rangle / 2aL = 1 - (1/x)(1 - e^{-x}) \tag{3}$$

$$3 \langle R^2 \rangle / aL = 1 - 3/x + (6/x^2)[1 - (1/x)(1 - e^{-x})] \tag{3'}$$

where $x = L/a$

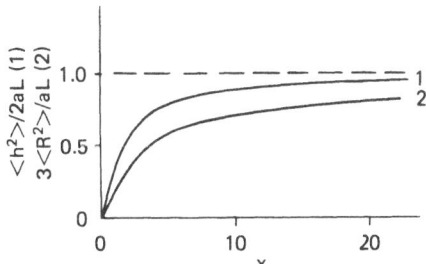

Fig. 1. Conformation of a persistent chain determined by the $\langle h^2 \rangle/2aL$ (1) and $3\langle R^2 \rangle/aL$ (2) ratios vs. its reduced length $x = L/a$[3, 4]

In Fig. 1 the values of $\langle h^2 \rangle/2aL$ vs. x (Curve 1) and $3\langle R^2 \rangle/aL$ vs. x (Curve 2) are plotted.

According to Eq. (3) at high x a worm-like chain becomes a Gaussian chain for which Eq. (1) (A = 2 a) is valid. At low values of x (x ⟶ 0) it follows from Eq. (3) that h ⟶ L, i. e. the worm-like chain adopts a straight conformation (Fig. 1). At a given value of a, a certain "degree of coiling" of the worm-like chain corresponds to each value of x whereas the value of a, just as that of A, serves as a quantitative measure of its equilibrium rigidity. Hence, a persistent chain is a universal model making it possible to describe the geometrical properties of chain molecules along the entire conceivable range of changes in their conformation from a straight rod to a Gaussian coil.

Quantitative information on the equilibrium flexibility of actually existing chain molecules permitting the determination of their A (or a) values can be obtained from the experimental study of their conformational, hydrodynamic, optical and other properties in dilute solutions[2, 4−9].

Extensive experimental findings obtained by using a "theta solvent"[5] reveal that for a vast majority of well-known chain polymers the length of the Kuhn segment A ranges from 15 to 30 Å[10].

These polymers are usually called flexible-chain polymers.

However, polymer chains are known for which A amounts to several hundred and even several thousand Ångströms[11, 12]. Such polymers are termed rigid-chain polymers.

According to the persistent model (Eq. (3)), the chain molecules exhibit a range of molecular weights M (range of changes in L and, hence, of changes in $x = 2 L/A$), in which the chain conformation is intermediate between a rod-like conformation and a Gaussian coil. This may be expressed by saying that the conformational properties of the molecule are those of a "semi-rigid chain". Evidently, the higher the equilibrium rigidity (i. e. A) of the chain, the higher the corresponding range of molecular weights. Thus, for common flexible-chain polymers for which A ≈ 15–30 A pronounced deviations from the Gaussian properties (in accordance with Eq. (3) and Fig. 1) begin only for oligomers, i. e. at M < 10⁴. In this case, the experimental determination of these deviations by direct measurements of $\langle h^2 \rangle$ is not possible and more sensitive methods are required, such as flow birefringence[13−16]. In contrast, for molecules with a rigidity corresponding to A ≈ 100 A, the departures from the Gaussian coil (decrease in $\langle h^2 \rangle/AL$ with M in a homologous series of polymer fractions) can be distinctly observed even at M ≈ 10⁵ and below by using direct measurements of $\langle h^2 \rangle$[17]. This circumstance is of major importance since it permits the separation of polymers into flexible-chain and rigid-chain polymers, the latter comprising polymers whose molecules of M > 10⁴ behave in solutions distinctly as "semi-rigid" chains (Fig. 1).

Usually, this corresponds to the values of A ⩾ 100 Å. The introduction of this criterion into equilibrium rigidity is confirmed experimentally since, as will be shown below, rigid-chain polymers classified in this manner exhibit some specific features markedly distinguishing them from flexible-chain polymers.

1.2 Chain Structure and Rigidity

The experimental data available permit to establish some structural features of polymer molecules leading to high chain rigidity.

The interaction of side groups is the main mechanism determining the equilibrium rigidity of typical flexible-chain polymers. This interaction is usually considered in the conformational statistics of polymer chains[5, 8]. However, this mechanism cannot lead to a considerable increase in the rigidity of the main chain. The investigations of "comb-like" molecules of poly(alkyl acrylate)s and poly(alkyl methacrylate)s[11] have revealed that, when the side chain is very long, the equilibrium rigidity of the chain cannot increase more than two- or threefold. This is quite insufficient to account for the properties characteristic of rigid-chain polymers.

The introduction of ring structures into the main chain is much more effective. It can increase the length of the Kuhn segment by more than an order of magnitude since cyclization drastically decreases or even prevents the intramolecular rotations about valence bonds.

Cellulose ethers and esters are well-known examples of rigid-chain polymers. The limited flexibility of their molecules is due to the presence of glucose rings and additional cyclic structures formed by hydrogen bonds of side groups[44] (Table 1).

The chains of polypeptides and polynucleotides in a helical conformation are even more "cyclized" by hydrogen bonds. They are characterized by very high equilibrium and kinetic rigidity[6].

Modern chemical synthesis permits the preparation of polymer molecules with a cyclo-linear structure in which cyclization is produced by covalent rather than by hydrogen bonds.

Ladder polymers with completely cyclized chains are examples of these molecules.

These polymers have been synthesized on the basis of polysiloxane macromolecules with a double chain structure and aromatic and aliphatic side groups[94–97] (Fig. 2). However, the length of the Kuhn segment for chains with a ladder structure can vary depending on the conditions of the synthesis (Table 2). This means that the defects in the ladder structure may play a certain part in the flexibility of these polymers. Nevertheless, the main mechanism of their flexibility involves the deformation of valence angles and bonds of their double-chain "network"

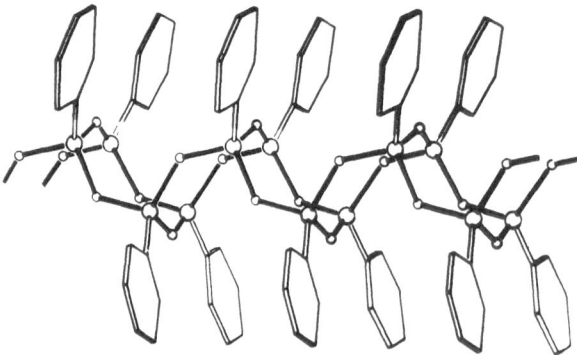

Fig. 2. Structure of ladder polyphenylsiloxane molecules

Table 1. Length of the Kuhn segment A and chain diameter d for cellulose esters and ethers according to the data of sedimentation – diffusion studies and viscometry

Polymer	Substituting group	Degree of substitution	Solvent	Sedimentation, diffusion		Viscometry		Ref.
				A (Å)	d (Å)	A (Å)	d (Å)	
1. Cellulose butyrate	$-O-CO-C_3H_7$	3.0	methyl ethyl ketone	260	10	260		42)
2. Cellulose benzoate	$-O-CO-C_6H_5$	2.2 ± 0.1	bromoform tetrachloroethane dioxane	120 240	5 20	130 130		44)
3. Cellulose carbanilate	$-O-CO-NH-C_6H_5$	2.2 ± 0.4	ethyl acetate	190	11			43)
4. Cellulose mono-phenyl acetate	$-O-CO-CH_2-C_6H_5$	2.7 ± 0.2	benzene	240	20			44)
5. Cellulose diphenyl phosphonocarbamate	$-O-CO-NH-\overset{O}{P}(OC_6H_5)_2$	2.4 ± 0.4	dioxane	240	46			44)
6. Cellulose nitrate	$-O-NO_2$	1.9	dioxane cyclohexanone methyl ethyl ketone	250	8	209 192	6.8 6.3	83)
7. Ethyl cellulose	$-O-C_2H_5$	2.5 ± 0.1	ethyl acetate	180	8			123)

Table 2. Length of the Kuhn segment A and chain diameter d for ladder polysiloxanes according to the data of sedimentation-diffusion studies and viscometry

Polymer	Substituents	Solvent	Sedimentation, diffusion		Viscometry	Ref.
			A (Å)	d (Å)	A (Å)	
1. Ladder polyphenyl-siloxane	$R^1 = R^2 = -C_6H_5$	benzene bromoform	200		158 178	28, 32, 33) 28)
2. Ladder polyphenyl-siloxane		benzene	136			30, 32, 33)
3. Ladder polyphenyl-siloxane		benzene	82		96 74	30, 32, 33) 30)
4. Ladder polyphenyl-siloxane		benzene	300	7.7	325	36)
5. Ladder poly(m-chloro-phenylsiloxane)	$R^1 = R^2 = -C_6H_4-Cl-m$	benzene	300	7	200	32, 33, 35)
6. Ladder poly(dichloro-phenylsiloxane)	$R^1 = R^2 = -C_6H_3Cl_2$	benzene benzene	250 220	17 14	180	37) 38)
7. Ladder poly(3-methyl--1-butenesiloxane)	$R^1 = R^2 = -CH=CH-CH(CH_3)_2$	butyl acetate	220		220	30−33)
8. Ladder poly(3-methyl-1-butenesiloxane)		butyl acetate	240	9	200 220	33, 34) 48)
9. Ladder poly(phenyl-isobutylsiloxane) (1:1)	$R^1 = -C_6H_5,\ R^2 = -CH_2-CH(CH_3)_2$	butyl acetate	96	9	100	33, 33, 41)
10. Ladder poly(phenyl-isohexylsiloxane) (1:1)	$R^1 = -C_6H_5,\ R^2 = -(CH_2)_3-CH(CH_3)_2$	butyl acetate	130	8.5	100	32, 33, 41, 48)
11. Linear poly(methyl-phenylsiloxane)	$R^1 = -CH_3,\ R^2 = -C_6H_5$	benzene			12.5	24, 32)

during its thermal vibrations. Evidently, this mechanism of flexibility greatly differs from that for linear polymers the chain of which becomes curved owing to internal rotations about valence bonds without the deformation of valence angles.

Conjugation in the chain can lead to even more rigid polymer structures. High resonance energy in the amide group[98] leads to its quasi-conjugation and coplanarity. Hence, the introduction of this group into the chain decreases the flexibility of the latter[99, 100]. However, in common polyamides the amide groups are separated in the main chain by methylene or aromatic groups ensuring some freedom of rotation and reducing chain rigidity to that of common flexible polymers[8, 101, 102]. The situation changes greatly if the amide groups in an aliphatic polyamide are very close to each other. This is the case with poly(alkyl isocyanate)s (nylon-1) whose molecules consist entirely of amide groups in which the hydrogen at the nitrogen atom is replaced with an alkyl group[103]. With a coplanar (*cis-* or *trans-*) configuration of each amide group this structure should lead to the coplanarity of all the bonds of the main chain of poly(alkyl isocyanate)s. Steric interactions of the group R with the neighboring carbonyl oxygen can cause a slight deviation from complete coplanarity of chain conformation but its structure retains regularity and a high degree of order: amide groups in *cis-* and *trans-*configurations[104, 105] alternate in the chain. The replacement of an aliphatic side group by an aromatic group (poly(tolyl isocyanate)) destroys conjugation in the chain and the resulting flexibility is similar to that of common polymers[110].

The synthesis of polymer molecules with chains containing both complex aromatic heterocycles and amide groups is being widely developed. Many of these molecules exhibit high equilibrium rigidity[106] and can form a nematic mesophase in concentrated solutions[107]. Hence, it is possible to use these polymers for the manufacture of thermally stable materials with a high modulus and good mechanical properties[108]. The parameter of equilibrium rigidity A for aromatic polymers may vary widely depending on the structural details of their chains. This will be considered in greater detail below.

We will also deal with some features of the behaviour of rigid-chain polymer molecules in dilute solutions with emphasis on their dinamooptical and electrooptical properties.

2 Hydrodynamic Properties of Rigid-Chain Polymer Molecules

2.1 Translational Friction and Viscosity

The study of hydrodynamic properties (sedimentation, diffusion and viscosity) of dilute polymer solutions is the most widely used method permitting the characterization of geometric properties (size and conformation) of polymer molecules.

To describe hydrodynamic properties of chain molecules, modern theories[9, 18−21] use the worm-like model characterized not only by the segment length A and the contour length L (Eq. 3) but also by the chain diameter d playing a major part in intramolecular hydrodynamic interactions[22, 23].

A high segment length value A of rigid chain polymers leads to many characteristic hydrodynamic properties of their solutions differing from those of flexible polymers.

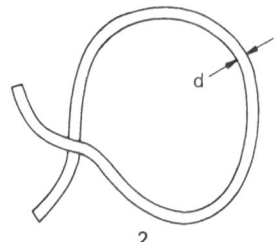

Fig. 3. Density of molecular coils
of flexible-chain (A/d is low, 1)
and rigid-chain (A/d is high, 2)
polymers at equal values of L and
d of the chain

For most rigid-chain polymers the ratio A/d is much higher than that for flexible chain polymers. Hence, the molecular coil of a rigid-chain polymer even when its chain length L (or M) in solution is large, is much more "loose" than a flexible-chain polymer (Fig. 3). As a result, these polymers exhibit two important properties.

Firstly, for rigid-chain polymers the excluded volume effects in thermodynamically good solvents (mainly determining the conformation and size of a flexible-chain molecule in solution[5, 9]) are negligible. This has been confirmed both experimentally[24] and theoretically[25].

Secondly, hydrodynamic interactions (depending on the d/A ratio) are weaker for rigid-chain coils than for flexible-chain coils and when their hydrodynamic parameters are considered, draining effect should be taken into account.

These peculiarities are included in the theories of viscosity and translational and rotational friction of worm-like chains[18-21, 26, 27] the results of which are shown in Eqs. (4) and (5)[26, 27]

$$f = \frac{kT}{D} = \frac{M (1 - \bar{v} \rho)}{N_A [s]} = P \eta_0 (LA)^{\frac{1}{2}} \tag{4}$$

$$[\eta] = \Phi (LA)^{\frac{3}{2}} / M \tag{5}$$

where $[\eta]$, D and $[s]$ are the intrinsic viscosity, translational diffusion and sedimentation coefficients of the polymer solution, f is the translational friction coefficient of the polymer molecule, M and $\bar{v}$ are the molecular weight and the partial specific volume of the polymer, η_0 and ρ are the viscosity and the density of the solvent, respectively, and P and Φ are functions of relative chain length L/A and of the parameter of hydrodynamic interaction, d/A, respectively. These functions have been represented in an analytical form and tabulated over a wide range of changes in the L/A and d/A parameters[26, 27]. At extremely high molecular weights (at L/A → ∞), functions P and Φ approach an asymptotic limit P ⟶ P_∞ = 5.11; Φ ⟶ Φ_∞ = 2.862 x 10^{23} [23, 27] (the Flory constant). This corresponds to the conformation of a hydrodynamically undrained Gaussian coil.

For the practical use of Eqs. (4) and (5) it is necessary to express the chain length, L, by an experimentally determined value of the molecular weight, M, according to

$$L = M\lambda/M_0 \tag{6}$$

where M_0 is the molecular weight of the monomer unit (the repeating unit) and λ the projection of this unit on the chain direction.

If we take into account Eq. (6), Eqs. (4) and (5) become

$$f = P\eta_0 M^{\frac{1}{2}} (\lambda A/M_0)^{\frac{1}{2}} \tag{4'}$$

$$[\eta] = \Phi M^{\frac{1}{2}} (\lambda A/M_0)^{\frac{3}{2}} \tag{5'}$$

Hence, the theory permits the determination of $\lambda A = S\lambda^2$ (where $S = A/\lambda$ is the number of monomer units in the Kuhn segment) by utilizing the experimental dependence of $[\eta]$ on M, or of f (i. e. D or [s]) on M). Consequently, to find the parameters of equilibrium chain rigidity, A or S, from hydrodynamic data, it is also necessary to know the value of λ. When this condition is fulfilled, the chain diameter, d, can also be determined.

For practical use, approximate expressions obtained from Eqs. (4) and (5) are suitable under certain conditions.

Thus, for relatively long $(L/A > 2.278)$ chains it follows from Eq. (4) that

$$\frac{\eta_0 DM}{kT} = \frac{N_A \eta_0}{1 - \bar{v}\rho} \quad [s] = \frac{1}{P_\infty} \cdot \left(\frac{M}{LA}\right)^{\frac{1}{2}} \cdot M^{\frac{1}{2}} + \frac{M}{3\pi L}\left(\ln \frac{A}{d} - 1.05\right) \tag{7}$$

Equation (7) describes the translational friction of a worm-like chain in the conformation of a "partially drained" coil. According to Eq. (7), the dependence of DM or [s] on $M^{\frac{1}{2}}$ for fractions is linear at relatively high values of M. This conclusion is confirmed by experimental data obtained for fractions of rigid-chain polymers: ladder polysiloxanes[24, 28–38, 41], polyisocyanates[39, 40], cellulose ethers and esters[42–44], aromatic polymers[45–47] and many other polymers (Figs. 4–7). For all the polymers investigated, the experimental points of [s] vs. $M^{0.5}$ fit straight lines at relatively high molecular weights. Their slope being equal to $[(1 - \bar{v}\rho)/N_A\eta_0 P_\infty] (M_0/\lambda A)^{0.5}$ according to Eq. (7) permits the determination of A if λ is known.

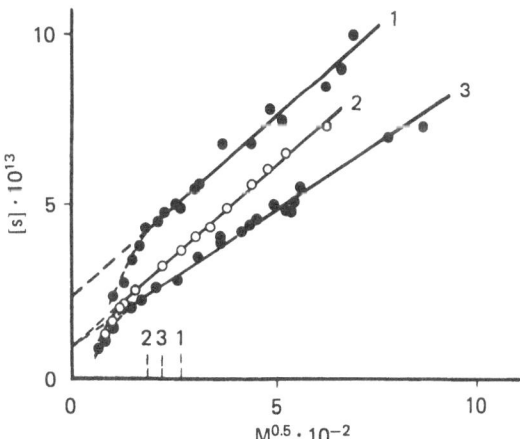

Fig. 4. Sedimentation coefficient [s] vs. molecular weight $M^{0.5}$ for ladder polysiloxanes: 1: poly-phenylsiloxane in benzene[36]; 2: poly(phenyl-isohexylsiloxane) (1:1) in butyl acetate[32, 33, 41]; 3: poly(3-methyl-1-butenesiloxane) (1:1) in butyl acetate[30–34]. Broken lines intersecting the abscissa correspond to the length of the main chain L = 2.27 $A^{26)}$

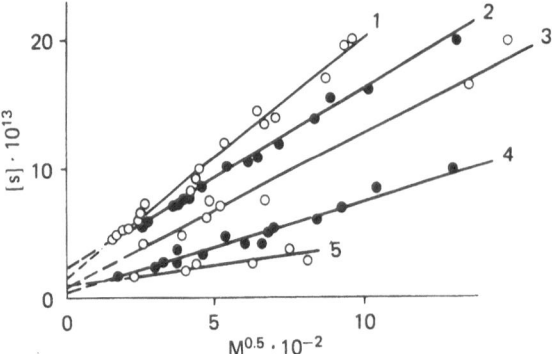

Fig. 5. Sedimentation coefficient [s] vs. molecular weight $M^{0.5}$ for cellulose esters: 1: cellulose carbanilate in ethyl acetate[43]; 2: cellulose butyrate in methyl ethyl ketone[42]; 3: cellulose monophenylacetate in benzene[44]; 4: cellulose diphenyl phosphonocarbamate in dioxane[44]; 5: cellulose benzoate in dioxane[44]

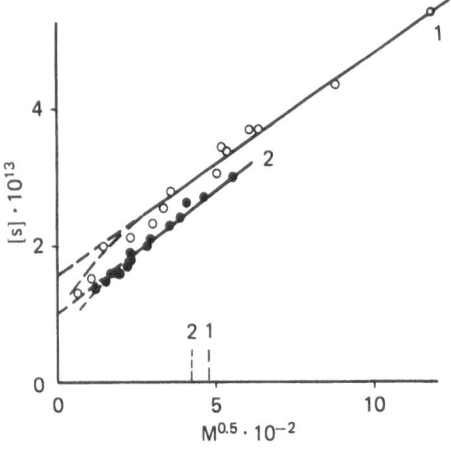

Fig. 6. Sedimentation coefficient [s] vs. molecular weight $M^{0.5}$ in tetrachloromethane: 1: poly(butyl isocyanate)[39]; 2: poly(chlorohexyl isocyanate)[40]. Broken lines intersecting the abscissa correspond to the main chain length $L = 2.27\ A$[26]

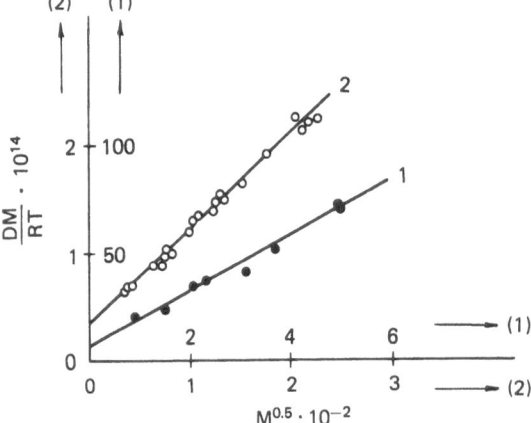

Fig. 7. DM/RT vs. molecular weight $M^{0.5}$ for aromatic polyamides: 1: poly(m-phenylene isophthalamide) in DMAA + 3% LiCl[45]; 2: poly(p-phenylene-1,3,4-oxadiazole) in sulfuric acid[47]

Extrapolating the straight part of the curve to $M \longrightarrow 0$ the value $\ln \left(\dfrac{A}{d}\right) - 1.05$ may be obtained from the intercept on the ordinate according to Eq. (6); thus the value of d is obtained.

The experimental values of A for some polymers found by means of this procedure are listed in Table 3.

Usually, the values of the hydrodynamic diameter, d, calculated from experimental data on sedimentation and diffusion correspond in their order of magnitude to transverse chain dimensions obtained from structural analysis. However, the precision of the determinations of d, using hydrodynamic data, is not high.

Figures 4 and 5 reveal that at low molecular weights the dependence of [s] on $M^{1/2}$ markedly deviates from a linear dependence. This is in agreement with the theory[12, 26] according to which Eq. (7) is not obeyed at $L/A < 2.278$. Eq. (7) should thus be replaced by another approximate expression, i.e.:

$$\frac{3 \pi \eta_0 DL}{kT} = \frac{3 \pi \eta_0 N_A \lambda}{M_0 (1 - \bar{v}\rho)} \qquad [s] = \ln \frac{L}{d} + 0.386 + 0.168 \frac{L}{A} \left(1 + 0.113 \frac{L}{A}\right) \tag{8}$$

in which the terms on the right-hand side containing L/A reflect the deviation of the molecular conformation from a straight rod due to chain flexibility.

When experimental data on the viscosity of a rigid-chain polymer are discussed, it is convenient to use Eq. (9) which follows from Eq. (5)[27] under the condition that $L/A < 2.278$ and $\dfrac{d}{A} \leqslant 0.1$

$$\frac{24 \, M \, [\eta]}{\pi N_A L^3} \left\{ \ln \frac{L}{d} - 1.839 + \frac{8.24}{\ln (L/d)} - \frac{32.86}{[\ln (L/d)]^2} + \frac{41.1}{[\ln (L/d)]^3} \right\} = f(x) \tag{9}$$

where

$$f(x) = \frac{24}{x^4} (e^{-x} - 1 + x - x^2/2 + x^3/6) =$$

$$= 1 - x/5 + x^2/30 - x^3/210 + x^4/1680 - \dots; \quad x = 2 \, L/A$$

For a thin straight (rod-like) chain ($x \longrightarrow 0$), the left-hand side of Eq. (9) is independent of L (or of M) and close to unity. Conversely, for a chain of finite rigidity when x (i.e. L or M) increases, the left-hand side of Eq. (9) decreases and this decrease is theoretically described by the function f(x).

For each actually existing polymer the left-hand side of Eq. (9) can be determined by using the experimental values of M and [η] (taking into account Eq. (6)) if the value of d is known even approximately. Studying the dependence of the left-hand side of Eq. (9) on M and comparing it with the function f(x) it is possible to determine A[48].

Figure 8 is an example of this plot for fractions of some rigid-chain polymers.

Equations (4) and (5) show that when the parameter $x = 2 \, L/A$ changes from 0 to ∞, the hydrodynamic properties of a worm-like chain change from those of a thin straight rod to those of an undrained Gaussian coil. In accordance with this the dependence of intrinsic viscosity [η] and diffusion coefficient D on molecular weight M of a rigid-chain polymer cannot be described by the usual Mark-Kuhn dependence

$$[\eta] = K_\eta M^\alpha, \quad D = K_D M^{-\beta} \tag{10}$$

over a wide range of changes in M with constant values of coefficients K_η, K_D, α, and β. When x changes from 0 to ∞, the value of α for a worm-like chain changes from 1.7 to 0.5 and that of β

Table 3. Length of the Kuhn segment A and chain diameter d for aliphatic and aromatic polyamides according to the data of sedimentation-diffusion studies and viscometry

Polymer	Monomer unit	Solvent	Sedimentation, diffusion		Viscometry	Ref.
			A (Å)	d (Å)	A (Å)	
1. Poly(butyl isocyanate)	$\cdots -N-CO-\cdots$ C_4H_9	carbon tetra chloride	1000	6	900	39, 48)
2. Poly(clorohexyl isocyanate)	$\cdots -N-CO-\cdots$ $C_6H_{12}Cl$	carbon tetra chloride	420	13	580	40, 48)
3. Poly(tolyl isocyanate)	$\cdots -N-CO-\cdots$ (CH_3 ring)	acetone	16		13	110)
4. Poly(m-phenylene isophthalamide)	$\cdots -NH-$ ... $-NH-CO-$ ... $-C-O-\cdots$	N,N-dimethylacetamide + 3% LiCl	47			45)
5. Poly(p-phenylene-1,3,4-oxadiazole)	(oxadiazole ring structure)	sulfuric acid	85	4.9		47)

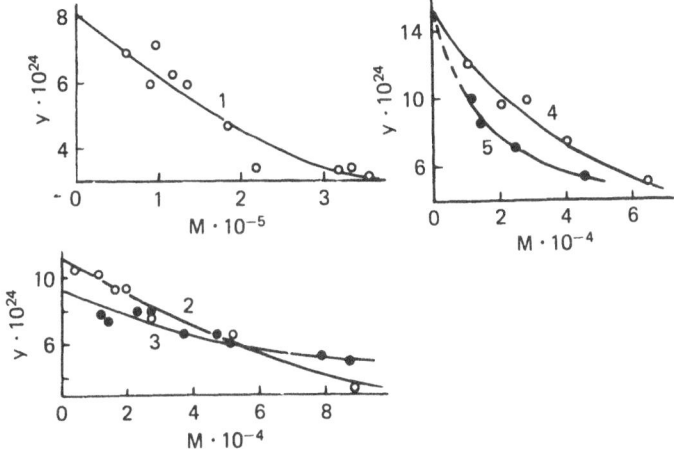

Fig. 8. Plot of y (left-hand side of Eq. (9)) vs. M for the following fractions: 1: poly(γ-benzyl-L-glutamate)[140]; 2: poly(butyl isocyanate)[39]; 3: poly(chlorohexyl isocyanate)[40]; 4: ladder poly (3-methyl-1-butenesiloxane)[34]; 5: ladder poly(phenyl-isohexylsiloxane) (1:1)[32]. Coincidence between the experimental points and the theoretical curve f (x) in Eq. (9) is attained at $\lambda/d = 1$ and at the following values of A: 1640 A (1), 900 A (2), 580 A (3), 220 A (4), and 100 A (5)[48]

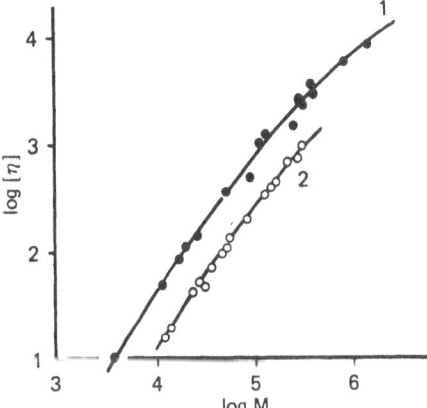

Fig. 9. Experimental dependence of log [η] on log M: 1: poly(butyl isocyanate)[39]; 2: poly(chlorohexyl isocyanate)[40] in tetrachloromethane

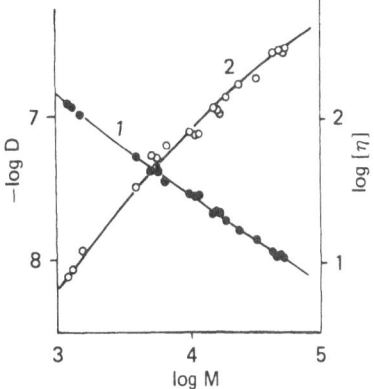

Fig. 10. Double logarithmic plot of diffusion coefficient D (1) and intrinsic viscosity [η] (2) vs. molecular weight M for poly (p-phenylene-1,3,4-oxadiazole) samples in sulfuric acid[47]. Curves 1 and 2 are theoretical curves[26, 27]

varies from 0.8 to 0.5. Experimental data confirm this conclusion as can be seen in Figs. 9 and 10 in which the dependence of $[\eta]$ and D on M is shown on a double logarithmic scale for poly(alkyl isocyanate) fractions and poly(p-phenylene-1,3,4-oxadiazole) samples. These dependences are represented by curves the slopes of which fall in these ranges[13].

As already mentioned, the experimental study of the dependence of $[\eta]$ on M or D on M allows the determination of A and d if the value of λ is known. It has been shown[27] that a combined investigation of the dependence of $[\eta]$ on M and D on M permits the calculation of all three structural characteristics of the polymer chain λ, A and d.

2.2 Some Approximate Methods for Evaluating Molecular Characteristics

Studying molecular properties of rigid-chain polymers by hydrodynamic methods, specific difficulties sometimes arise. Thus, many polymers with aromatic chains that are of great practical importance are molecularly soluble only in very aggressive media such as concentrated sulfuric acid. Hence, experiments in these systems require specific instruments[49, 50].

Another and greater difficulty in the investigation of many rigid-chain polymers very important in practice is the impossibility of applying such classical method of conformational analysis as sedimentation in an ultracentrifuge. It is therefore natural to take into account low molecular weights M of many of these polymers and weak hydrodynamic interactions in their chains (low d/A, Fig. 3). Since the sedimentation coefficient[5] of a polymer molecule is proportional to M/f the values of [s] for a rigid-chain polymer (at equal values of L, d and density in the same solvent) are lower than for a flexible-chain polymer. Moreover, in a theoretically limiting case of a "completely drained" chain molecule (without hydrodynamic interactions) f is proportional to M and, hence, [s] is independent of molecular weight. The greater the chain rigidity (i.e. A/d), the closer approach the hydrodynamic properties that limit and the weaker is the dependence of [s] on M. The situation is even more unfavorable because the use of sulfuric acid as a solvent is required. Its viscosity is higher by an order of magnitude than that of common organic solvents and the value of [s] is lower by the same factor. Consequently, both the absolute value of [s] and its dependence on M are lower than the resolution power of high-speed sedimentation used in practice.

The diffusion method is much more effective for studying polymers with high equilibrium flexibility. Since the translational diffusion coefficient D of a molecule in solution depends only on the friction coefficient f and is inversely proportional to the latter the dependence on M is the stronger the higher the rigidity of chain molecules (in contrast to the dependence of [s] on M). For very rigid-chain polymers at low M this dependence is close to inverse proportionality (Fig. 10) and can be used in the study of molecular characteristics.

In the determination of the molecular weight of the polymer, sedimentation investigations of its solutions may be replaced by viscometric measurements. It is known[51−53] that a combination of intrinsic viscosity, $[\eta]$, and diffusion coefficient, D, leads to the equation[51−53]

$$\eta_0 D(M[\eta])^{\frac{1}{3}}/T \equiv A_0 \tag{11}$$

where parameter A_0 for flexible-chain polymers is virtually independent of M and its average experimental value is $A_0 = 3.4 \times 10^{-10}$ erg/deg (if $[\eta]$ is expressed in dl/g).

For molecules represented by a worm-like chain, the parameter A_0 can be calculated theoretically since the combination of Eqs. (4) and (5) gives

$$A_0 = k \, \Phi^{\frac{1}{3}} P^{-1} \tag{12}$$

i.e., A_0 is a function of L/A and d/A.

Calculations reveal[47] that for worm-like chains A_0 changes relatively slightly with L/A and d/A whereas for polymers with high chain rigidity its value is greater than for flexible-chain polymers. This corresponds to experimental data according to which the average experimental value of A_0 for fractions of rigid-chain poly(alkyl isocyanate)s is $3.9 - 4 \times 10^{-10}$ erg/deg[39, 40] and for poly(m-phenylene isophthalamide)[49] and poly(amide hydrazide)[46] it is $A_0 \approx 3.7 \times 10^{-10}$ erg/deg.

These results permit the determination of molecular weights of rigid-chain polymers from experimental values of D and $[\eta]$ by using Eq. (11) and the values of $A_0 = 3.8 \times 10^{-10}$ erg/deg which correspond to P = 5.11 and $\Phi = 2.862 \times 10^{21}$ ($[\eta]$ is measured in dl/g). The values of M calculated in this manner[54] are reasonable and agree with the results obtained by direct methods (light scattering)[55].

Tables 1--3 give the values of molecular parameters characterizing the equilibrium rigidity and hydrodynamic properties of molecules of some rigid-chain polymers obtained from viscometric and diffusion-sedimentation measurements of their dilute solutions.

2.3 Rotational Friction

Rotational friction is an important property of macromolecules and plays a significant role in such phenomena as viscosity and birefringence of solutions in mechanical or electric fields.

In a laminar flow or in an external potential field a polymer molecule is subjected to forces that can both make it rotate as a whole and cause a relative shift of its parts leading to a deformation, i.e. changing its conformation. Which of these two mechanisms of motion predominates depends on the ratio of times required for the deformation and rotation of the molecule. If the time of the rotation of the molecule as a whole, τ_0, is shorter than the time required for its deformation, τ_d, its rotation as a whole will be the main type of motion. In the process considered here this molecule may be called kinetically rigid. In the opposite case, when $\tau_0 > \tau_d$, the deformation mechanism of motion will predominate and the molecule will be kinetically flexible. To characterize quantitatively the kinetic rigidity of chain molecules Kuhn has introduced[30] the concept of "internal viscosity" — a quantity describing the resistance of the molecule to a rapid change in its shape. Later, the theory of internal viscosity has been developed by Cerf [65, 81].

In principle, the study of the motion of the macromolecule by the action of rotating forces can be used for characterizing its kinetic rigidity.

However, viscometric measurements of dilute polymer solutions in a steady flow are inadequate for this purpose although, as already indicated, viscosity is related to molecular rotation. This has been demonstrated by Zimm's theory[56]. Zimm considered the kinetics of the motion and deformation of a kinetically flexible polymer chain in a weak mechanical field with harmonic velocity gradient g at frequency ν. It has been found that under steady and weak flow conditions

($\nu \longrightarrow 0$, $g \longrightarrow 0$), although each chain segment participates in random thermal motion, it exhibits rotational motion with respect to the center of gravity of the molecule. The mean angular velocity of this motion is $g/2$, just as for a kinetically rigid spherically symmetrical chain model[57]. This motion is the source of losses due to friction. Hence, the intrinsic viscosity of a solution of kinetically flexible Gaussian chains in a steady flow does not differ from that of a solution of kinetically rigid molecules frozen in the conformation of a Gaussian coil. Differences arise only in the unsteady (sinusoidal) flow and are detected in the relaxation spectra. In accordance with this, modern theories of hydrodynamic properties of polymer molecules and related transfer processes[18, 19, 26, 27] discussed before (p. 104), consider "average" conformations of rigid-chain models and only their large-scale motion disregarding the kinetics of intramolecular (small-scale) motions.

The main characteristics determining the rotational mobility of a rigid particle (molecule) are its rotational friction coefficient, W, depending on the size and shape (conformation) of the molecule, the viscosity of the solvent η_0 and the position of the the axis of rotation.

Precise expressions for W have been obtained for the model of a rigid ellipsoid of revolution (spheroid) rotating about the central axis normal to the symmetry axis of the particle[58, 59].

$$W = v \eta_0 f(p) \tag{13}$$

where v is the volume of the ellipsoid, p its axial ratio (shape asymmetry) and function $f_0(p)$ at $p > 1$ (elongated ellipsoid) is given by

$$f_0(p) = 4 \frac{p^4 - 1}{p^2} \left(\frac{2 p^2 - 1}{2 p \sqrt{p^2 - 1}} \cdot \ln \frac{p + \sqrt{p^2 - 1}}{p - \sqrt{p^2 - 1}} - 1 \right)^{-1} \tag{14}$$

The theory of the viscosity of solutions containing ellipsoidal particles also relates the value of $[\eta]$ to the volume, v, the molecular weight M and the degree of the asymmetry p of the particle

$$[\eta] = N_A v \nu(p)/M \tag{15}$$

where N_A is Avogadro's number. Function $\nu(p)$ is 2.5 for a sphere ($p = 1$)[60] and for $p > 1$ it can be approximated by[2]

$$\nu(p) = 2.5 + 0.4075 (p - 1)^{1.508} \tag{16}$$

at $1 < p < 15$ and

$$\nu(p) = 1.6 + \frac{p^2}{5} \left[\frac{1}{3 (\ln 2 p - 1.5)} + \frac{1}{\ln 2 p - 0.5} \right] \tag{17}$$

at $p > 15$.

Comparison of Eqs. (13) and (15) gives

$$\eta_0 [\eta] M = F(p) N_A W \tag{18}$$

where function $F(p) = \nu(p)/f_0(p)$ is determined by Eqs. (14), (16) and (17). It follows from these equations that, when p changes from ∞ (a rod-like molecule, a very elongated ellipsoid) to 1 (a sphere), F (p) increases from F (∞) = 2/15 to F (1) = 5/12.

The analysis of Eqs. (14), (16) and (17) reveals[61] that in the range of changes in p from ∞ to 2.5 they are equivalent to the linear relationship

$$F(p) = 2/15 + 0.382514/p \tag{19}$$

Hence, Eq. (18) shows that for a solution of ellipsoidal particles a very general relationship exists between the intrinsic viscosity and the rotational friction coefficient of the particle.

A similar relationship exists for a solution in which the hydrodynamic properties of a dissolved molecule are approximated by a rigid-chain necklace model of spherical beads.

Thus, if the necklace has the shape of a rigid straight rod, its rotational friction coefficient for the rotation about the central axis normal to the rod is given by

$$W = N \zeta \langle R^2 \rangle \tag{20}$$

Here $\langle R^2 \rangle$ is the mean-square radius of gyration of the molecule, N the number of beads in the chain and ζ the average effective translational friction coefficient of a bead, depending on the hydrodynamic interaction in the chain. For a straight necklace $\zeta = \zeta_0 / \ln N$ [62, 63] where ζ_0 is the friction coefficient of a bead in the absence of hydrodynamic interactions. The sign $\langle \rangle$ designates averaging over all chain conformations (for a straight rod their number is unity).

The intrinsic viscosity of a solution of molecules described by a kinetically rigid chain necklace that can adopt any conformation (from the straight rod to the spherically symmetrical bead distribution) is given by [23, 57]

$$[\eta] = (1/6) \zeta N_A \langle R^2 \rangle / M_0 \eta_0 \tag{21}$$

where $M_0 = M/N$ is the molecular weight of the bead and ζ the effective friction coefficient depending on chain conformation and hydrodynamic interaction. For a straight rod ζ in Eq. (21) has the same value as in Eq. (20).

A comparison of Eqs. (20) and (21) for a straight necklace yields

$$\eta_0 [\eta] M = N_A \langle W \rangle / 6 \tag{22}$$

independent of the hydrodynamic interaction in the chain. When the distribution of the centers of hydrodynamic resistance (beads) in a chain molecule is spherically symmetrical, its rotational friction coefficient with respect to any axis passing through the center of symmetry is given by

$$\langle W \rangle = \frac{2}{3} \zeta N \langle R^2 \rangle \tag{23}$$

where ζ assumes a value characteristic of a spherical conformation.

A comparison of Eq. (23) with Eq. (21) in which the value of ζ for a spherical conformation is the same as in Eq. (23) for a spherically symmetrical necklace yields the equation

$$\eta_0 [\eta] M = N_A \langle W \rangle / 4 \tag{24}$$

Equation (24) has been obtained[23] for a Gaussian chain since the Gaussian coil is considered as a spherically symmetrical cloud of segments.

Hence, when the average conformation of a chain molecule changes from the limiting asymmetrical (rod-like with the axial ratio p = ∞) to the limiting symmetrical (spherical, p = 1) conformation, the form of the relationship between $[\eta]$ M and W is retained but the numerical factor changes from 1/6 to 1/4, according to Eqs. (22) and (24).

If a worm-like chain with the conformation varying from a straight rod to a Gaussian coil (when the parameter x = 2 L/A varies from 0 to ∞) is used for the description of the hydrodynamic properties of the molecule, then the following equation is also valid for it

$$\eta_0 [\eta] M = F_x N_A \langle W \rangle \tag{18'}$$

where the coefficient F_x is the function of x increasing with x from $F_0 = 1/6$ (at x = 0) to F_∞ (at x $\longrightarrow \infty$) that depends on the shape asymmetry p_∞ exhibited by the Gaussian coil.

Rotational friction of a kinetically rigid worm-like chain has been considered by Hearst[19]. The position of the centers of hydrodynamic resistance (beads) in the worm-like model used by Hearst is determined in system of XYZ coordinates. Its origin coincides with the middle point of the chain and the direction of the Z axis coincides with the chain direction at this point (Fig. 11). It is assumed that the distribution of chain elements (beads) is cylindrically symmetrical with an axis of symmetry Z. In a molecular system of XYZ coordinates this distribution is given by

$$\langle z_i \rangle = AL_i/3 - (A^2/18)(1 - e^{-6\ L_i/A}) \tag{25}$$

$$\langle x_i \rangle = \langle y_i \rangle = AL_i/3 - (2/9)\ A^2 + (A^2 e^{-2\ L_i/A})/4 - (A^2 e^{-6\ L_i/A})/36 \ldots \tag{26}$$

where $\langle x_i \rangle$, $\langle y_i \rangle$ and $\langle z_i \rangle$ are averaged (over all conformations) squares of the coordinates of the ith chain element located at distance L_i from the origin along the chain contour (Fig. 11).

Taking into account the hydrodynamic interactions between the elements of a worm-like chain, as in the calculation of translation friction and visosity[18, 62, 63], Hearst[19] has obtained an expression for the coefficient of rotatory diffusion D_r of the molecule with respect to the central axis normal to its axis of symmetry (axis X or Y in Fig. 11). For the limiting cases of a short and long worm-like chain this expression is given by

$$\frac{D_r}{kT} = \frac{3}{\pi \eta_0 L^3} \left[\left(\ln \frac{L}{d} - 1 \right) + 0.75 \frac{L}{A} \left(\ln \frac{L}{d} - 2.07 \right) \right] \tag{27}$$

for the conformation of a slightly bent rod and by

$$\frac{\eta_0 D_r M^2}{kT} = 0.72 \left[\left(\frac{M}{LA} \right)^{\frac{3}{2}} M^{\frac{1}{2}} + \left(\frac{M}{L} \right)^2 A^{-1} \cdot 0.884 \left(\ln \frac{A}{d} - 1.43 \right) \right] \tag{28}$$

for the conformation of a worm-like coil.

The first term on the right-hand side of Eq. (27) corresponds to the expression for D_r of a thin straight rod[62, 63] and the second term containing L/A expresses the deviation from a straight conformation caused by finite flexibility of the chain.

The first term in Eq. (28) corresponds to the hydrodynamic properties of an undrained Gaussian coil and the second term describes the draining effect caused by the "loose" structure of a worm-like coil (high value of A/d).

Equations (27) and (28) can be used in experimental investigations of the rotational mobility of macromolecules by means of electric birefringence as described below.

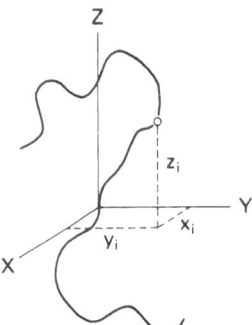

Fig. 11. Coordinate system of a worm-like chain. X, Y and Z are ortogonal axes with the origin at the mid-point of the chain. Axis Z is tangential to the chain contour at the coordinate origin. x_i, y_i and z_i are coordinates of the i-th chain element

The molecular model shown in Fig. 11 was also been employed in the theory of the viscosity of solutions of rigid-chain polymers[20]. The results of this theory are shown as expressions for $[\eta]$ for the limiting cases of a slightly bent rod and a worm-like coil. A comparison of these expressions with Eqs. (27) and (28) reveals that Eq. (22) for a rod-like conformation and Eq. (24) for a Gaussian coil are also fulfilled for this model. Hence, at a relatively large chain length $(L/A \gg 1)$ rotational friction of the worm-like model described in a system of XYZ coordinates (Fig. 11) corresponds to the hydrodynamic properties of a spherically symmetrical spatial distribution of segments.

This result is due to the specificity of the molecular system of coordinates fixing axes with respect to which the rotational friction coefficient is calculated.

The asymmetry of the shape of the worm-like chain leading to the difference in friction coefficients for the rotation about the X (or Y) axis and about the Z axis (Fig. 11) according to Hearst[19] is determined by the relation $\sum_i \langle z_i^2 \rangle / \sum_i \langle y_i^2 \rangle$ where $\langle z_i^2 \rangle$ and $\langle y_i^2 \rangle$ are expressed by Eqs. (25) and (26) and summation is carried out over all chain elements. This summation gives

$$p_x^2 = \sum_i \langle z_i^2 \rangle / \sum_i \langle y_i^2 \rangle = \frac{x^2 - (2/3)\, x + 2/9 - (2/9)\, e^{-3x}}{x^2 - (8/3)\, x + 26/9 - 3\, e^{-x} + (1/9)\, e^{-3x}} \tag{29}$$

where x, as in Eq. (3), characterizes the relative chain length $x = 2\, L/A$.

It follows from Eq. (29) that at $x \longrightarrow 0$, $p_x^2 \longrightarrow 4/x \longrightarrow \infty$, corresponds to an infinitely thin straight rod. In the Gaussian coil conformation x tends to infinity and it follows from Eq. (29) that p_x^2 tends to unity, corresponding to a model with a spherically symmetrical segment distribution.

3 Flow Birefringence

Flow birefringence is a widely used method for the investigation of optical, conformational and hydrodynamic properties of molecules of flexible-chain polymers in solution. The experimental and theoretical material available is very comprehensive and has been considered in many reviews[2, 4, 6, 64-70] providing the basis for the understanding of this phenomenon not only qualitatively but also quantitatively.

In the field of rigid-chain polymers the situation is much less favorable. Relatively few experimental findings have been reported. Probably, this is associated with the experimental difficulties involved and the lack of polymer samples available for studies since a greater progress in the synthesis of rigid-chain macromolecules with a well-known chemical structure has been made only during the last decade. The number of papers on the dynamo-optical properties of rigid-chain molecules is even smaller and most publications only deal with the elucidation of individual phenomena and do not give a full description of all experimentally detected properties and facts from the standpoint of a single theory.

However, the analysis of experimental data on flow birefringence and the use of some theoretical (sometimes semi-empirical) concepts allows important conclusions to be drawn on optical, conformational and structural characteristics of rigid-chain molecules[11, 12]. Hence, we may consider the method of birefringence to be an effective method for studying the structure of rigid-chain polymers, particularly in those cases where other methods fail. Probably, this consideration may serve as a justification for writing the present review.

Comparisons of dynamo-optical properties of rigid- and flexible-chain polymers lead to the important conclusion that usually polymer molecules with high equilibrium chain rigidity are also characterized by a high kinetic rigidity (Fig. 12).

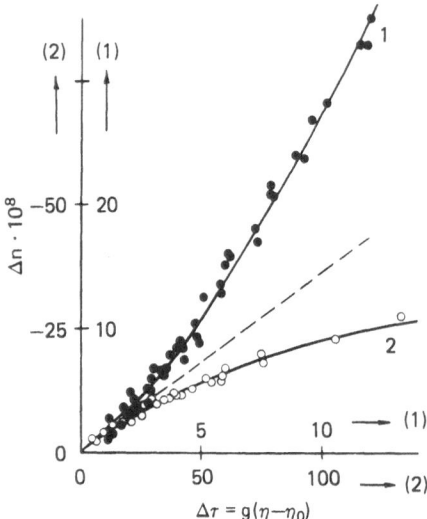

Fig. 12. Plot of flow birefringence Δn vs. shearing stress $\Delta \tau = g(\eta - \eta_0)$ for polyisobutylene in benzene (M = 132000, the range of c is $0.84 - 0.28$ g/dl)[71] (1) and cellulose nitrate in cyclohexanone (M = 500000, degree of substitution = 2.8; the range of c is $0.15 - 0.03$ g/dl)[83] (2)

Curve 1 describes the dependence of flow birefringence, Δn, on effective shearing stress, $g(\eta - \eta_0)$ (g is the velocity gradient, η and η_0 are the viscosities of solution and solvent, respectively) for solutions of polyisobutylene[71], a typical flexible-chain polymer (the length of the Kuhn segment is 20 Å[5]). The curve exhibits a pronounced upward curvature indicating that the deformation (stretching) of the coil under the influence of shearing stress in flow plays an important role in the observed birefringence. This deformation effect is well known for all flexible-chain polymers and can lead to a very specific dependence of Δn on g including the change in the sign of the anisotropy of solution[67].

Curve 2 in Fig. 12 reveals a similar dependence of Δn on $g(\eta - \eta_0)$ for solutions of cellulose nitrate[83], a typical rigid-chain polymer (the length of the Kuhn segment is 200 Å – see Table 1). Curve 2 has a downward curvature corresponding to the dependence predicted by the theory of rigid (non-deformed) chains of an asymmetrical shape[4]. This means that the conformation of cellulose nitrate molecules (and of other rigid-chain polymers) in a shearing flow does not undergo such great changes as that of polyisobutylene and its flow birefringence in solution is caused to a considerable extent by a non-uniform rotation of a coil optically anisotropic and asymmetric in shape.

Hence, to understand correctly the mechanism of flow birefringence of rigid-chain polymers it is necessary to take into account both the asymmetry of the shape of their molecules and their optical anisotropy. Both these properties are the main properties of the molecule and if one of them is absent, flow birefringence in solution of kinetically rigid molecules is impossible.

3.1 Asymmetry of the Shape of Chain Molecules

The asymmetry of the shape of a worm-like chain in the conformation of a straight rod is uniquely determined by the $p = L/d$ ratio whereas the asymmetry of the shape of a chain molecule in the Gaussian coil conformation can be characterized in various ways.

Even in his early works Kuhn has shown that the most probable form of a random Gaussian coil is described by a non-spherical distribution of segments[1].

A direct experimental proof of the asymmetry of the coil shape exhibited in flow is the well-known optical macro-form effect[71] comprehensively investigated in solutions of flexible-chain polymers[67].

It is possible to evaluate theoretically the asymmetry of the shape of a Gaussian coil by calculating the distances between a pair of chain elements (fixed in each conformation) located at the greatest distance from each other in two mutually perpendicular directions[72, 73]. This method yields the average asymmetry of the coil shape $p = 2$.

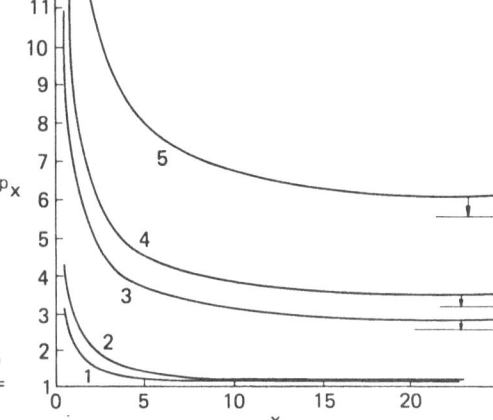

Fig. 13. Shape asymmetry p of a worm-like chain vs. $x = 2L/A$. 1: curve according to Eq. (29). 2–5: curves according to Eq. (32): 2: $p_\infty = 1$; 3: $p_\infty = 2.5$; 4: $p_\infty = 3.1$; 5: $p_\infty = 5.5$

Another method of characterizing the asymmetry of the molecular shape is related to the choice of the main direction in the molecule. In the principle axis Z in Fig. 11 tangential to the worm-like chain at its middle point can be chosen as this direction. However, in this system of coordinates the asymmetry of the shape of the molecule is adequately expressed only in the range of conformations close to a rod whereas for the Gaussian coil conformation $(x \longrightarrow \infty)$, according to Eq. (29), p is equal to unity and the coil is spherical (see also Fig. 13). Hence, the asymmetry of the shape of the Gaussian coil cannot be adequately expressed by a system of coordinates determined by the middle element of a worm-like chain.

In another characterization of the main direction in a chain molecule according to Kuhn, vector h, joining two ends of the chain is chosen for this purpose[1]. In this evaluation of the degree of asymmetry of a Gaussian coil, length h can be compared with transverse dimensions of the coil determined by distances h_1 and h_2 from vector h relative to the chain points situated at the greatest distance from h in two mutually perpendicular directions normal to h[72]. It has been found that $\langle h^2 \rangle : \langle h_1^2 \rangle : \langle h_2^2 \rangle = 1 : (1/6) : (1/24)$. Correspondingly, the square of the degree of asymmetry of the Gaussian coil is given by

$$p^2 = 2 \langle h^2 \rangle / (\langle h_1^2 \rangle + \langle h_2^2 \rangle) = 48/5 \quad \text{and} \quad p = 3.1$$

Recently, in a number of theoretical computational studies[74-78] the shape of the Gaussian coil has been characterized by the components of radii of gyration R_1, R_2 and R_3 in the three main directions of the coil (fixed for each conformation). Calculations reveal that the average shape of the coil can be approximated by a three-axial ellipsoid with the ratio of squares of axes $\langle R_3^2 \rangle : \langle R_2^2 \rangle : \langle R_1^2 \rangle = 11.7 : 2.7 : 1$. Hence,

$$p = [2 \langle R_3^2 \rangle / (\langle R_2^2 \rangle + \langle R_1^2 \rangle)]^{0.5} = 2.5$$

The asphericity of instantaneous conformations of chain molecules is also demonstrated by experiments with tagged polymers[79].

Investigating the effect of the macroform in the birefringence phenomenon[67], the order of magnitude of the theoretical values of p is in good agreement with that of experimental values obtained for some flexible-chain polymers.

These data suggest that the average shape of the Gaussian coil is far from being spherical and it is possible to establish its greatest asymmetry by characterizing it in a system of coordinates in which the vector of the end-to-end distance of the chain is the axis of cylindrical symmetry.

In the same system of coordinates the asymmetry of the shape of a worm-like chain may also be evaluated over the entire range of possible conformations from a thin straight rod to a Gaussian coil[82, 61].

For this purpose, a worm-like coil is represented by an extended body of revolution for which the central radius of gyration $\langle R^2 \rangle$ is related to the longitudinal, H, and transverse, Q, body dimensions by the general equation

$$\langle R^2 \rangle = \gamma H^2 + \delta Q^2 \tag{30}$$

where γ and δ are constant coefficients depending on the model chosen. Thus, for an ellipsoid of revolution $\gamma = 1/20$; $\delta = 1/10$ and for a cylinder $\gamma = 1/12$ and $\delta = 1/8$. Evidently, the asymmetry of the shape of a molecular model is $p = H/Q$.

Equation (30) is supplemented by the following conditions: 1) at all values of $x = 2$ L/A Eqs. (3) and (3') characteristic of a worm-like chain are obeyed for $\langle h^2 \rangle$, $\langle R^2 \rangle$ and x; 2) at all values of x the long axis of the coil H is equal to $\alpha \langle h^2 \rangle^{1/2}$ where α is the constant coefficient; 3) at $x \longrightarrow 0$ the asymmetry of the shape of a worm-like chain p_x tends to infinity (i.e., the chain diameter, d, is relatively small) and, according to Eqs.(3) and (3'), the ratio $\langle R^2 \rangle / \langle h^2 \rangle$ tends to 1/12; 4) at $x \longrightarrow \infty$, p_x tends to p_∞ and, according to Eqs. (3) and (3'), the ratio $\langle R^2 \rangle / \langle h^2 \rangle$ approaches 1/6. Using these conditions and Eq. (30)

$$1/p_x^2 = (12/p_\infty^2) (\langle R^2 \rangle / \langle h^2 \rangle - 1/12) \tag{31}$$

If Eqs. (3) and (3') are taken into account, Eq. (31) is equivalent to

$$\frac{1}{p_x^2} = \frac{6}{p_\infty^2} \left[\frac{1/3 - 1/x}{1 - (1 - e^{-x})/x} + \frac{2}{x^2} - \frac{1}{6} \right] \tag{32}$$

Equation (32) uniquely relates the asymmetry of the shape of a worm-like chain p_x to the relative chain length x and the asymmetry of the chain shape p_∞ in the conformation of a Gaussian coil.

The dependence of p_x on x according to Eq. (32) is described in Fig. 13 (Curves 2–5) at various hypothetical values of p_∞. Curve 1 reveals the same dependence according to Eq. (29).

Curve 1 is close to Curve 2 corresponding to $p_\infty = 1$ although both curves are obtained in different systems of coordinates: Curve 1 is obtained in the central system (Fig. 11) and Curve 2 in a system related to vector h. In both cases at $x \longrightarrow 0$, p tends to infinity, which corresponds to the conformation of a straight infinitely thin rod. As the chain length increases, the asymmetry markedly decreases and even for chains containing 2–3 Kuhn segments it attains virtually an asymptotic limit corresponding to a spherically symmetrical distribution of segments.

Curves 3, 4 and 5 also show a sharp decrease in the asymmetry of the shape of the molecule with increasing length although in these cases an asymmetrical Gaussian coil is the asymptotic limit. According to experimental data and theoretical studies, Curve 3 (the asymptotic limit of which $p_\infty = 2.5$ corresponds to recent theoretical data) probably exhibits the closest correspondence to the dependence of p on x for the existing rigid-chain polymers.

3.2 Optical Anisotropy of a Worm-Like Chain

As already mentioned, the optical anisotropy of a rigid-chain molecule is its second main property determining the value and sign of birefringence observed in solution. It is characterized by the difference between the main optical polarizabilities γ_1, γ_2 and γ_3 determined in a system of coordinates related to the chain.

3.2.1 In Coordinate Systems Related to a Chain Element

A symmetrical cylindrical system may be chosen as the axes of coordinates. The symmetry axis of this system should coincide with the direction of some (e.g. the first) element of a worm-like chain[82, 84] (Fig. 14).

Each element of the worm-like chain of length ΔL is characterized by the axial symmetry of its optical properties and its anisotropy is βL where β is the optical anisotropy of the unit length of a worm-like chain.

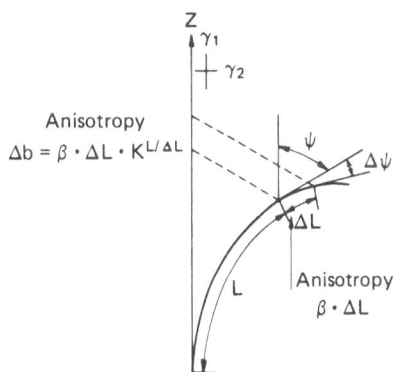

Fig. 14. Optical anisotropy of a worm-like chain in a system of coordinates of its first element

Chain element ΔL located at distance L from the chain end along the contour leads to the average effective anisotropy in the axes of the first element

$$\Delta b = \beta \cdot \Delta L \cdot K^{L/\Delta L} \tag{33}$$

Here the factor

$$K \equiv (1/2)\, (3\, \langle \cos^2 \Delta \Psi \rangle - 1) \tag{34}$$

depending on the curvature of the chain characterizes the decrease in the effective anisotropy $(-\Delta b/\Delta L)$ of the unit length of the chain with a displacement along its contour at distance ΔL. This distance corresponds to change $\Delta \Psi$ in angle Ψ formed by the chain and the direction of its first element. In general, brackets $\langle\rangle$ refer to the averaged overall chain conformations.

Equation (33) yields the differential form of the dependence of the effective chain anisotropy, b, on chain length L.

$$\frac{db}{dL} = \beta K^{L/\Delta L} = \beta\, (e^{\ln K})^{\frac{L}{\Delta L}} = \beta e^{\left(\frac{\ln K}{\Delta L}\right) L}$$

and, correspondingly, the integral form is given by

$$b = \beta\, (\Delta L/\ln K) \left[e^{\left(\frac{\ln K}{\Delta L}\right) L} - 1 \right] \tag{35}$$

When the worm-like chain is long, it becomes a Gaussian chain and, according to Eq. (35), its anisotropy is given by

$$\lim_{L \to \infty} b \equiv b_\infty = -\beta\, (\Delta L/\ln K) \tag{36}$$

Hence

$$b = b_\infty\, (1 - e^{-\beta L/b_\infty}) \tag{37}$$

On the other hand, the curvature of the chain can be expressed by its persistent length a according to Eq. (2) (see p. 4):

$$\ln \langle \cos \Delta \Psi \rangle = -\Delta L/a \tag{38}$$

Comparison of Eqs. (36) and (38) gives

$$b_\infty = a\beta \ln \langle \cos \Delta \Psi \rangle / \ln K$$

and, taking into account that

$$\lim_{\Delta \Psi \to 0}\, [\ln \langle \cos \Delta \Psi \rangle / \ln K] = 1/3$$

we obtain $b_\infty = (1/3)\, a\beta$

Finally, the substitution of this equation into Eq. (37) gives

$$\gamma_1 - \gamma_2 = (1/3)\, a\beta\, (1 - e^{-3x}) =$$
$$= \beta L\, [1 - (3/2)\, x + (3/2)\, x^2 - (9/8)\, x^3 + \ldots] \tag{39}$$

where $x = L/a$ and γ_1 and γ_2 are optical polarizabilities of a worm-like chain parallel and perpendicular to the Z axis (Fig. 14), respectively.

Analogously to Eq. (39), the anisotropy of a worm-like chain of length L in a system of coordinates of its middle element (Fig. 11) is given by

$$\gamma_1 - \gamma_2 = \gamma_z - \gamma_x = \gamma_z - \gamma_y = \frac{2}{3} \, a\beta \left(1 - e^{-\frac{3x}{2}} \right) =$$

$$= \beta L \left(1 - \frac{3}{4} x + \frac{3}{4} x^2 - \ldots \right) \tag{40}$$

Equations (39) and (40) reveal that the anisotropy of a worm-like chain in the limiting conformation of a straight rod (at $x \longrightarrow 0$) is equal to βL, i.e. it is proportional to the contour chain length regardless of whether it is calculated in a system of coordinates of the first or the middle chain element.

In the limiting case of a Gaussian coil ($x \longrightarrow \infty$) the anisotropy of a worm-like chain in a system of coordinates of its middle element is given by

$$(\gamma_1 - \gamma_2)_\infty = (2/3) \, a\beta = (1/3) \, A\beta = (1/3) \, (\alpha_1 - \alpha_2)$$

i.e., it is equal to one third of the optical anisotropy of the Kuhn segment[2, 80]. In a system of coordinates of the end element it is twice as small.

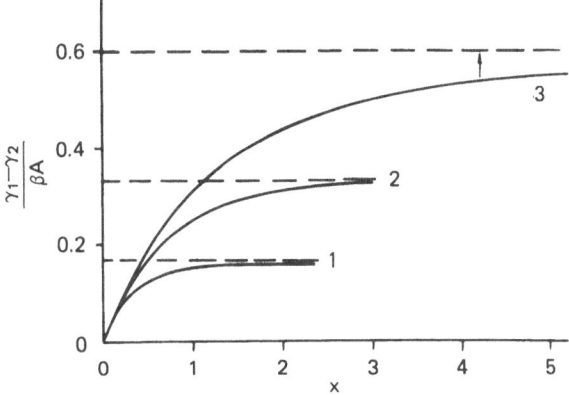

Fig. 15. Plot of relative anisotropy of a worm-like chain $\gamma_1 - \gamma_2/\beta A$ vs. relative chain length $x = 2 \, L/A$; 1: In a system of coordinates of the first chain element in accordance with Eq. (39); 2: In a system of coordinates of the middle chain element according to Eq. (40), 3: In a system of coordinates of vector h according to Eq. (47)

Figure 15 shows the dependence of the chain anisotropy (with respect to the anisotropy of the Kuhn segment βA) on its relative length $x = L/a$ according to Eqs. (39) and (40). At $x \longrightarrow 0$ the curves have the slope 0.5 and rapidly attain an asymptotic limit corresponding to the anisotropy of the Gaussian chain.

Equation (39) is employed for the characterization of the optical properties of comb-like molecules (see below). However, Eqs. (39) and (40) cannot be utilized for the description of the flow birefringence of kinetically rigid linear chains since in the system of molecular coordinates to which they belong the Gaussian coil is, on the average, spherically symmetrical (see p. 115).

3.2.2 In Coordinate Systems Related to Chain Ends

A system of molecular coordinates related to vector h joining the chain ends is more useful for studying the phenomenon of flow birefringence (Fig. 16).

As a result of the chain structure of a polymer molecule a preferred orientation of the chain elements exists in the direction of vector h. If ϑ is the angle formed by chain element ΔL and direction h (Fig. 16), it follows that for any chain conformation corresponding to a given value of h the value of $\overline{\cos \vartheta}$ averaged over all chain elements is given by

$$\overline{\cos \vartheta} = h/L \tag{41}$$

This means that for any finite h the value of $\overline{\cos \vartheta}$ is non-zero and, hence, direction h is the direction of the preferred orientation of chain elements.

If each chain element ΔL is optically anisotropic and uniaxially symmetrical, it introduces the following anisotropy with respect to the h axis

$$\Delta \gamma = \beta L (3 \cos^2 \vartheta - 1)/2 \tag{42}$$

To find the anisotropy of the whole molecule in molecular coordinates h, the value of $\cos^2 \vartheta$ should be averaged over all chain elements and all its conformations corresponding to a given value of h. To solve this problem Kuhn and Grün[85] have considered the statistics of the intramolecular orientational distribution of segments with respect to h in a freely jointed chain and have shown that for the most probable distribution the value of $\overline{\cos^2 \vartheta}$ averaged over all chain segments can be expressed as a unique function of the h/L ratio

$$\overline{\cos^2 \vartheta} = 1 - 2 (h/L)/\mathscr{L}^* (h/L) \tag{43}$$

where $\mathscr{L}^* (h/L)$ is the "inverse" Langevin function determined by the series

$$\mathscr{L}^* (h/L) = 3 \, h/L + \frac{9}{5} \left(\frac{h}{L}\right)^3 + \frac{297}{175} \left(\frac{h}{L}\right)^5 + \frac{1539}{875} \left(\frac{h}{L}\right)^7 + \ldots$$

Replacement of $\cos^2 \vartheta$ in Eq. (42) by its average value from Eq. (43) gives the difference between the polarizabilities of a freely jointed chain along the h axis and normal to it

$$\gamma_1 - \gamma_2 = \beta L \left[1 - \frac{3 \, h}{L} / \mathscr{L}^* \left(\frac{h}{L}\right) \right] \tag{44}$$

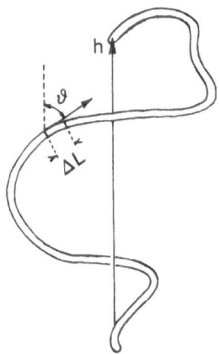

Fig. 16. Orientation of a chain element ΔL with respect to vector h

It can be easily shown[86] that a good approximation of $\mathcal{L}^*\left(\dfrac{h}{L}\right)$ is given by

$$\mathcal{L}^*\left(\frac{h}{L}\right) = \frac{3h}{L}\frac{(h/L)^2}{1 - (2/5)\,(h/L)^2}$$

Hence, substitution of this expression into Eq. (44) yields

$$\gamma_1 - \gamma_2 = \frac{3}{5}\,\beta L\,\frac{(h/L)^2}{1 - (2/5)\,(h/L)^2} \tag{45}$$

Equation (45), just as Eq. (44), is obeyed at all values of h : $0 \leqslant h \leqslant L$, i.e. it is valid for chain molecules at any degree of coiling h/L and in any conformation from a completely straight rod to a Gaussian coil. The mechanism ensuring the chain end-to-end distance h has not been taken into account in the derivation of these equations. Hence, it may be resonably assumed that Eqs. (44) and (45) can be applied not only to flexible-chain molecules but also to worm-like chains for which a small degree of coiling is determined by the skeletal rigidity, limiting the thermal mobility of chain elements, rather than by the external stretching force.

On this assumption the anisotropy of a worm-like chain with contour length L and persistent length a averaged over all conformations can be determined from Eq. (45) if in this equation the value of h^2 means the value averaged over all possible conformations of the chain in its equilibrium state, i. e. $\langle h^2 \rangle$

$$\langle \gamma_1 - \gamma_2 \rangle = \frac{3}{5}\,\beta L\,\frac{\langle h^2 \rangle / L^2}{1 - (2/5)\,\langle h^2 \rangle / L^2} \tag{46}$$

The substitution of $\langle h^2 \rangle$ according to Eq. (3) transforms Eq. (46) into Eq. (47):

$$\langle \gamma_1 - \gamma_2 \rangle = \frac{3}{5}\,\beta A f(x) \equiv \frac{6}{5}\,\beta L\,\frac{f(x)}{x} =$$

$$= \beta L\left(1 - \frac{5}{9}\,x + \frac{85}{324}\,x^2 - \frac{170}{1458}\,x^3 + \ldots\right) \tag{47}$$

where

$$f(x) = \frac{1 - (1 - e^{-x})/x}{1 - (0.8/x)\,[1 - (1 - e^{-x})/x]} \tag{48}$$

It follows from Eqs. (46)–(48) that the optical anisotropy of a worm-like chain in a system of h coordinates at low x changes proportionally to βL to attain an asymptotic limit of $(3/5)\beta A$ in the Gaussian range. The dependence of $\langle \gamma_1 - \gamma_2 \rangle / \beta A$ on x is shown by Curve 3 in Fig. 15 according to Eq. (47).

A comparison of Curves 1, 2 and 3 in Fig. 15 reveals that in the calculation of the anisotropy of the worm-like chain the use of different systems of molecular coordinates does not change the general character of the dependence of $\langle \gamma_1 - \gamma_2 \rangle$ on x. However, the maximum value of the asymptotic limit is obtained when vector h is used as the main direction in the chain molecule.

3.2.3 Form Anisotropy

The difference $\langle \gamma_1 - \gamma_2 \rangle$ between the polarizabilities of a worm-like chain depends not only on the values of L and A determining its conformation but also directly on the β parameter determined by the optical properties of the structural elements of a chain molecule. The value of β is related to the difference in the main polarizabilities $\Delta a \equiv a_\parallel - a_\perp$ of the monomer unit of the polymer in directions parallel ($a_\parallel$) and perpendicular ($a_\perp$) to the chain length, respectively

$$\beta = \Delta a / \lambda \qquad (49)$$

where λ is the length of the monomer unit in the chain direction. The anisotropy of the monomer unit, Δa, of the real chain molecule in the solvent medium is governed not only by the character of the polarizability of atomic groups in the unit but also depends on the refractive index of the solvent owing to the "microform" effect[67].

The higher the equilibrium chain rigidity and the increment of refractive index dn/dc in the polymer-solvent system, the more pronounced is this effect. For a rigid-chain polymer the Δa value in Eq. (49) is calculated according to[67]

$$\Delta a = \Delta a_i + \frac{M_0}{2 \pi N_A \bar{v}} \left(\frac{dn}{dc} \right)^2 \qquad (50)$$

in which the first term, Δa_i, corresponds to the intrinsic anisotropy of the unit and the second term containing the molecular weight of the monomer unit, M_0, and the partial specific volume of the polymer, $\bar{v}$, determines the anisotropy due to the microform effect. In solutions of flexible polymers the "macroform" effect can play an important role; it is proportional to M^2/v where v is the volume of the coil in solution[67]. However, for rigid-chain molecules for which v is much greater than for flexible-chain molecules (Fig. 3) this effect is small and may be neglected.

3.3 Characteristic Value of Flow Birefringence[61]

3.3.1 Molecular Model

Taking into account the foregoing considerations, the most suitable molecular model for describing the phenomenon of the flow birefringence of a rigid-chain polymer in solution appears to be a worm-like chain with the axis of symmetry of hydrodynamic and optical properties directed along the end-to-end vector h. It is assumed that the molecules exhibit relatively high kinetic rigidity, i.e. that the conformation of the chain remains unchanged during the time interval which is large compared with that required for the establishment of a steady orientational distribution of molecules in solution.

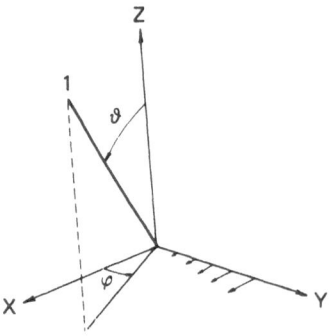

Fig. 17. Molecule with an axial symmetry of optical and hydrodynamic properties in a laminar flow. X: flow direction; Y: direction of velocity gradient; Z: direction of observation; 1: axis of symmetry of the molecule

The position of this molecule in flow is determined by angles ϑ and φ in the system of coordinates shown in Fig. 17.

The asymmetry of the shape of a rigid molecule accounts for its non-uniform rotation in the velocity-gradient field g resulting in the kinematic orientation of the molecular axes.

The function of the orientational distribution of molecules $\rho\,(\varphi, \vartheta)$ is found by solving the equation of rotatory diffusion in a steady laminar flow[19, 20]

$$\frac{1}{\sin\vartheta} \frac{\partial}{\partial\vartheta}\left(\sin\vartheta\, \frac{\partial\rho}{\partial\vartheta}\right) + \frac{1}{\sin^2\vartheta} \frac{\partial^2\rho}{\partial\varphi^2} = \frac{3}{2}\frac{gb}{D_r}\sin^2\vartheta \cdot \sin 2\varphi \tag{51}$$

where the asymmetry parameters of the shape of the molecule are given by

$$b = (p^2 - 1)/(p^2 + 1) \tag{52}$$

where p is determined according to Eq. (29) or Eq. (32), depending on the choice of axis 1 (Fig. 17) in the molecule; D_r is the coefficient of rotatory diffusion of the molecule with respect to the axis normal to the axis of molecular symmetry. D_r is related to the rotational friction coefficient W by

$$D_r = kT/W \tag{53}$$

The solution of Eq. (51) at low values of g/D_r taking into account Eq. (53) leads to the expression for $\rho\,(\varphi, \vartheta)$

$$\rho\,(\varphi, \vartheta) = \frac{1}{4\,\pi}\,[1 + \{(gbW)/4\,kT\}\sin 2\varphi \cdot \sin^2\vartheta + \ldots] \tag{54}$$

Equation (54) provides the basis for the calculation of the flow birefringence of a polymer solution at low g. Using Eq. (54) the characteristic value of birefringence may be obtained by a well-known method[4] according to

$$[n] \equiv \lim_{\substack{g \to 0 \\ c \to 0}} \left(\frac{\Delta n}{g\eta_0 c}\right) = \frac{2\,\pi\,N_A}{135\,M\,\eta_0\,kT}\,\frac{(n^2 + 2)^2}{n}\,\langle bW\,(\gamma_1 - \gamma_2)\rangle \tag{55}$$

where Δn is the measured value of birefringence, i.e. the difference between two main refractive indices of a solution at concentration c with refractive index n at velocity gradient g, M is the molecular weight of the polymer and η_0 the viscosity of the solvent.

It should be noted that Eq. (55) contains a factor b which becomes zero when the molecule is spherical (p = 1). Hence, the contribution to the birefringence of a solution is provided only by molecules adopting conformations in which the molecule is not only optically anisotropic but also geometrically aspherical.

3.3.2 Conformational Polydispersity

As before, brackets $\langle\rangle$ in Eq. (55) refer to averaging over all chain conformations. This averaging is necessary because the conformations of various kinetically rigid molecules in solution, at a given moment, vary for different molecules and each of them is characterized by certain values of $\gamma_1 - \gamma_2$, W and b, although they are virtually frozen.

In the first approximation b is considered to be the factor depending on the "average" conformation and the product $b \langle W (\gamma_1 - \gamma_2) \rangle$ can be used.

Equations (20) and (23) show that the average rotational friction coefficient W is proportional to the radius of gyration R^2 both for extended and symmetrical conformations. Hence, it is proportional to h^2 (compare Eqs. (3) and (3') p. 98). According to Eq. (46), optical anisotropy in the random chain conformation is also proportional to h^2 in the Gaussian range ($h/L \ll 1$). Hence, a Gaussian coil obeys the following equation

$$\langle W (\gamma_1 - \gamma_2) \rangle / (\langle W \rangle \langle \gamma_1 - \gamma_2 \rangle) = \langle h^4 \rangle / \langle h^2 \rangle^2 \tag{56}$$

According to Eq. (56), $\langle W (\gamma_1 - \gamma_2) \rangle / (\langle W \rangle \langle \gamma_1 - \gamma_2 \rangle)_{coil} = 5/3$. Only one rod-like conformation is possible for a worm-like chain at $x \longrightarrow 0$; hence, for a rod-like chain $\langle W (\gamma_1 - \gamma_2) \rangle / (\langle W \rangle \langle \gamma_1 - \gamma_2 \rangle)_{rod}$ is unity but $\langle h^4 \rangle / \langle h^2 \rangle^2$ is also unity. Thus, for a worm-like chain Eq. (56) is fulfilled for the possible limiting values: $x = 0$ and $x = \infty$. It is natural to assume that Eq. (56) is also obeyed at all intermediate values of x for a worm-like chain for which[3, 87]

$$\frac{\langle h^4 \rangle}{\langle h^2 \rangle^2} = \frac{\dfrac{5}{3} - \dfrac{52}{9\,x} - \dfrac{2}{27\,x^2}\,(1 - e^{-3\,x}) + \dfrac{8}{x^2}\,(1 - e^{-x}) - \dfrac{2}{x}\,e^{-x}}{[1 - (1 - e^{-x})/x]^2} \tag{57}$$

Hence, taking into account the above considerations and applying Eqs. (55) and (56) we obtain

$$\langle n \rangle = [2 \pi N_A (n^2 + 2)^2 / 135 \, M \eta_0 n K T] \, b \, \langle W \rangle \langle \gamma_1 - \gamma_2 \rangle \, \langle h^4 \rangle / \langle h^2 \rangle^2 \tag{58}$$

where $\langle h^4 \rangle / \langle h^2 \rangle^2$ is determined according to Eq. (57).

3.3.3 Dependence on Chain Length

The rotational friction coefficient $\langle W \rangle$ in Eq. (58) can be expressed by the intrinsic viscosity of solution $[\eta]$. Using Eq. (18') we obtain

$$[n]/[\eta] = B \, b \, (F_0/F_x) \langle \gamma_1 - \gamma_2 \rangle \langle h^4 \rangle / \langle h^2 \rangle^2 \tag{59}$$

where $F_0 = 1/6$ is the value of coefficient F_x in Eq. (18') at $x = 0$. Coefficient B is given by

$$B \equiv 4 \pi (n^2 + 2)^2 / 45 \, kTn \tag{59'}$$

Just as for an ellipsoid (Eq. (19)), for a worm-like chain coefficient $F_x = F_x (p)$ depends on the asymmetry of its shape p_x determined by Eq. (32) and changes with x. However, this change is less drastic than for an ellipsoid and ranges from $F_x (\infty) = 1/6$ to $F_x (1) = 1/4$. Hence, it is possible to find the value of the F_x coefficient corresponding to a worm-like chain with the asymmetry of the shape p_x from the value of F (p) for an ellipsoid (Eq. 19) with the same asymmetry $p = p_x$ according to

$$\{F_x (p_x) - F_x (\infty)\} / \{F (p) - F (\infty)\} = \{F_x (1) - F_x (\infty)\} / \{F (1) - F (\infty)\}$$

At $2.5 \leqslant p_x \leqslant \infty$ and taking into account Eq. (19) this equation is equivalent to

$$F_x (p_x) = 1/6 + 0.112504/p_x \tag{60}$$

where p_x is governed by Eq. (32).

Hence, when the values of p_x, β and A are given, all values on the right-hand side of Eq. (59) can be expressed according to Eqs. (52), (60), (47) and (57) as a function of the parameter $x = 2 L/A$.

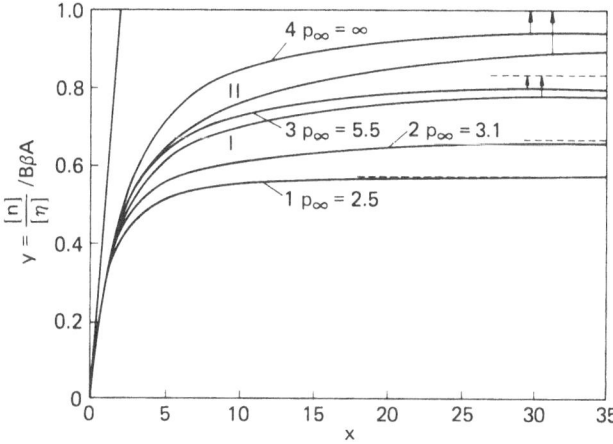

Fig. 18. $([n]/[\eta])/B\beta A$ vs. x plot for a worm-like chain at different values of the shape asymmetry p_∞ of the Gaussian coil: 1: $p_\infty = 2.5$, $y_\infty = 0.570$; 2: $p_\infty = 3.1$, $y_\infty = 0.665$; 3: $p_\infty = 5.5$, $y_\infty = 0.883$; 4: $p_\infty = \infty$, $y_\infty = 1$. I: according to the theory in Ref.[88]; II: according to the theory in Ref.[91]

In Fig. 18 $y \equiv ([n]/[\eta])/B\beta A$ is plotted vs. x according to Eq. (59) at various assumed values of the parameter of the asymmetry of the shape of a Gaussian coil p_∞.

All the curves reveal that at low x the ratio $[n]/[\eta]$ increases proportionally to x and subsequently y approaches a constant value, y_∞, corresponding to dynamo-optical properties of a kinetically rigid Gaussian coil. The higher the assumed value of the asymmetry of the shape of a worm-like chain in the Gaussian range, p_∞, the higher is the limiting value of y_∞. The lower the chosen value of p_∞, the more rapidly the curve approaches y_∞ (at lower x). These relationships are due to the fact that the dependence of $[n]/[\eta]$ on x is determined by a combination of two processes: an increase in the optical anisotropy $\langle \gamma_1 - \gamma_2 \rangle$ of a worm-like chain with growing length (according to Eq. (47)) and a simultaneous lowering of the factor bF_0/F_x owing to a decrease in the shape asymmetry of a worm-like coil (according to Eq. (32)). Curve 4 corresponds to a constant value of the factor $bF_0/F_x = 1$ in the entire range of changes in x. Hence, the dependence represented by Curve 4 does not take into account the change in the shape of a worm-like coil with increasing chain length.

Curve I describes the dependence of y on x obtained for kinetically rigid worm-like chains[88, 89] disregarding the dependence of p_x on x. The asymptotic limit of Curve I is 0.833 ... (instead of $y_\infty = 1$ for Curve 4) because in Reference[88] the anisotropy of the Gaussian chains is taken to be $\beta A/2$[90] rather than $3\beta A/5$ used for plotting Curves 1–4.

Figure 18 shows also Curve II corresponding to the Noda-Hearst theory[91] for kinetically flexible chains in which the local rigidity is taken into account on the basis of the Hearst-Harris dynamic model[92]. Factor B in this theory is equal to 3/5 of the value of B determined by Eq. (59'). In the Gaussian range ($x \longrightarrow \infty$) the Noda-Hearst theory[91] is completely equivalent to Zimm's theory[56] for kinetically flexible chain molecules. According to this, the limiting value of y_∞ for Cuve II is unity. Comparison of Curve II with Curves 1–4 reveals that the former resembles most closely Curve 4. This is due to the fact that in both cases a "kinetic unit" is used with the asymmetry of shape $p = \infty$. For Curve 4 this is the entire molecule and for Curve II it is a chain part representing its local rigidity. All the curves in Fig. 18 exhibit a common initial slope equal to 1/2 and demonstrate that the general character of the dependences of $[n]/[\eta]$ on x is the same.

However, both the steepness of the curve and the limiting value greatly depend on the molecular model used.

Table 4. Values of $y \equiv ([n]/[\eta])/B\beta A$ vs. x according to Eq. (59) at $p_\infty = \infty$ and $p_\infty = 2.5$

x	0	0.5	1	2	3	4	5	6	8	10	15	20	∞
$p_\infty = \infty$ y	0	0.196	0.322	0.483	0.583	0.655	0.706	0.749	0.803	0.840	0.891	0.918	1.000
$p_\infty = 2.5$ y	0	0.180	0.283	0.397	0.455	0.490	0.510	0.528	0.544	0.553	0.563	0.566	0.570

In Table 4 are listed the values of $y \equiv ([n]/[\eta])/B\beta A$ vs. x corresponding to Curve 1 ($p_\infty = 2.5$) and Curve 4 ($p_\infty = \infty$, b $F_0/F_x = 1$) in Fig. 18.

3.4 Characteristic Orientation Angle

Another characteristic of flow birefringence apart from the [n] value is the orientation angle α formed by the optical axis of an anisotropic solution and the flow direction. The characteristic value of the orientation angle is the value determined at extremely low shear rates

$$\left[\frac{\chi}{g} \right] \equiv \lim_{g \to 0, c \to 0} (\chi/g)$$

where $\chi = \pi/4 - \alpha$. To calculate this value it is necessary to solve diffusion Eq. (51) and to find the distribution function (54) taking into account the quadratic term with respect to g.

For a solution of kinetically rigid particles (molecules) identical in size and shape, the characteristic orientation angle is uniquely related to coefficients of rotatory diffusion D_r and rotational friction W of a particle

$$\left[\frac{\chi}{g} \right] = \frac{1}{12 D_r} = \frac{W}{12 \, kT} \tag{61}$$

Taking into account Eq. (18) Eq. (61) becomes

$$\left[\frac{\chi}{g} \right] = G \frac{\eta_0 \, [\eta] \, M}{RT} \tag{62}$$

where $G = 1/12 \, F (p)$ $\hspace{3cm}$ (62')

For ellipsoidal particles coefficient G depends only on the asymmetry p of their shape. When the shape changes from a rod (p = ∞) to a sphere (p = 1), G decreases from $G_{rod} = 0.625$ to $G_{sphere} = 0.2$ (Fig. 19, Curve 1)[14, 16, 17].

Equation (62) can be proved theoretically also for chain molecules but in this case the value of G depends on the model used.

Thus, for kinetically flexible chain molecules in the conformation of a Gaussian coil the value of G depends on coil draining and for the two limiting cases of weak and strong hydrodynamic interactions it is 0.2 and 0.1, respectively[56].

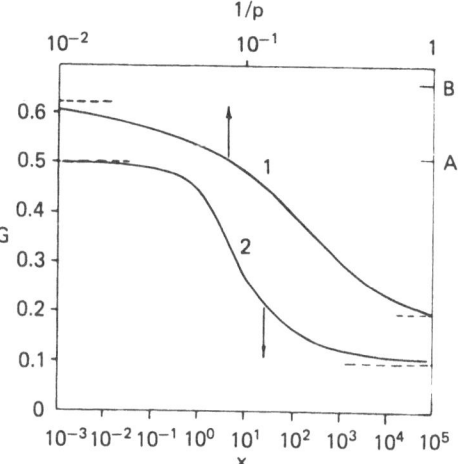

Fig. 19. Coefficient G as a function of molecular shape. 1: G vs. degree of shape asymmetry p of ellipsoidal particles; 2: G vs. x = 2 L/a for kinetically flexible chains with local rigidity according to the theory in Ref. [91]. A, B = values of G for kinetically rigid Gaussian coils at strong and weak hydrodynamic interactions, respectively [93]

The Noda-Hearst theory[91] calculates the G value for kinetically flexible chains over the entire range of possible changes in their conformations from a straight rod (x = 0) to a Gaussian coil (x = ∞). The corresponding curve describing the dependence G (x) for strong hydrodynamic interactions is shown in Fig. 19 (Curve 2). The limiting value of G in the Gaussian range (x ⟶ ∞) corresponds to Zimm's theory[56]; for a rod (x = 0, straight necklace) G is 0.5. The dependences of G on 1/p for ellipsoids (Curve 1) and on x for kinetically flexible chain molecules (Curve 2) are roughly similar. They show that coefficient G becomes lower with increasing compactness of the molecular structure (decrease in shape asymmetry and increase in hydrodynamic interactions).

In principle, for kinetically rigid chain molecules, just as for ellipsoids, Eq. (62) is obtained as a consequence of Eq. (61) since in this case the value and orientation of flow birefringence are determined by the rotational mobility of a molecule as a whole. However, for an assembly of chains of similar masses but "frozen" in different conformations it is necessary to take into account the conformational distribution.

The calculation of orientation angles for kinetically rigid Gaussian coils yields G = 0.667 and G = 0.504 for weak and strong hydrodynamic interactions, respectively[93]. These values exceed three- to five-fold the corresponding theoretical values of G for a kinetically flexible Gaussian coil[56, 91, 65, 80]. These high values of G (and, hence, of [χ/g] at equal values of [η] and M) for kinetically rigid coils as compared to those of kinetically flexible coils can be caused by the effect of conformational polydispersity. In fact, rigid chains with extended conformations are oriented at lower g values than coiled rigid chains. This should lead to the increase in the initial slope of the curve of the dependence of χ on g in the assembly of a conformationally polydisperse rigid chain.

In contrast, if rigid chain molecules are of a rod-like shape (for a worm-like model x ⟶ 0), there is no conformational polydispersity in the assembly and comparison of Eqs. (61) and (62) gives G = 0.5. This is close to the value of G for a kinetically rigid Gaussian coil both for weak (G = 0.667) and strong (G = 0.504) hydrodynamic interactions.

Hence, in contrast to flexible-chain molecules, when the length of kineticall rigid chains increases and their conformation changes from a straight conformation to a Gaussian coil (x varies from 0 to ∞ in a worm-like model), the theoretical value of G in Eq. (62) is approximately constant remaining close to 0.5. Presumably, this is a result of the compensation for the decrease in G caused by the increase in hydrodynamic interactions with rising x by its simultaneous increase due to growing conformational polydispersity.

4 Flow Birefringence. Experimental Data

4.1 Dependence on Molecular Weight

4.1.1 General Form of Dependence

In a comparison study of the values of flow birefringence Δn and the viscosity η of a polymer solution it is often possible to simplify the experimental procedure so as to avoid the determinations of the characteristic values of [n] and [η] by determining the quantity $\Delta n/g\,(\eta - \eta_0)$ at finite solution concentration instead of the ratio [n]/[η]. Here η_0 is the viscosity of the solvent and the value of $g(\eta - \eta_0) \equiv \Delta\tau$ characterizes the effective shearing stress in solution introduced by the dissolved polymer. Many experimental data show that for a flexible-chain polymer in the absence of the macroform effect the ratio $\Delta n/\Delta\tau$, which may be called the "shear optical coefficient", is independent of solution concentration, and, also, over a wide range of molecular weights, of chain length[67].

For rigid-chain polymers $\Delta n/\Delta\tau$ is also independent of solution concentration. This can be seen in Fig. 12 in which the experimental poonts for solutions of the same polymer at different concentrations fall on the same curve illustrating the dependence of Δn on $\Delta\tau$. This coincidence holds for both flexible-chain polyisobutylene (Curve 1) and rigid-chain nitrocellulose (Curve 2).

However, in contrast to flexible chain polymers, for rigid-chain polymers, the value of $\Delta n/\Delta\tau$ = [n]/[η] changes with chain length (molecular weight). This is demonstrated by Figs. 20–22 for various samples of ladder polymers, cellulose ethers and esters and polyisocyanates.

For all these polymers [n]/[η] is plotted virtually only at relatively high molecular weights but this ratio sharply decreases with M and approaches zero at M $\longrightarrow$ 0. This type of the dependence of [n]/[η] on M revealed experimentally for rigid-chain polymers qualitatively resembles the theoretical curves in Fig. 18. This is demon-

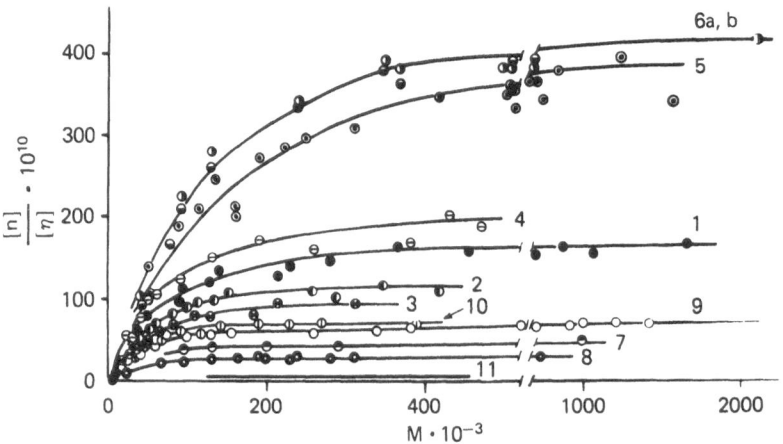

Fig. 20. Shear optical coefficient [n]/[η] as a function of molecular weight M for solutions of various samples of ladder polysiloxanes. Numbers on curves correspond to those in Table 5

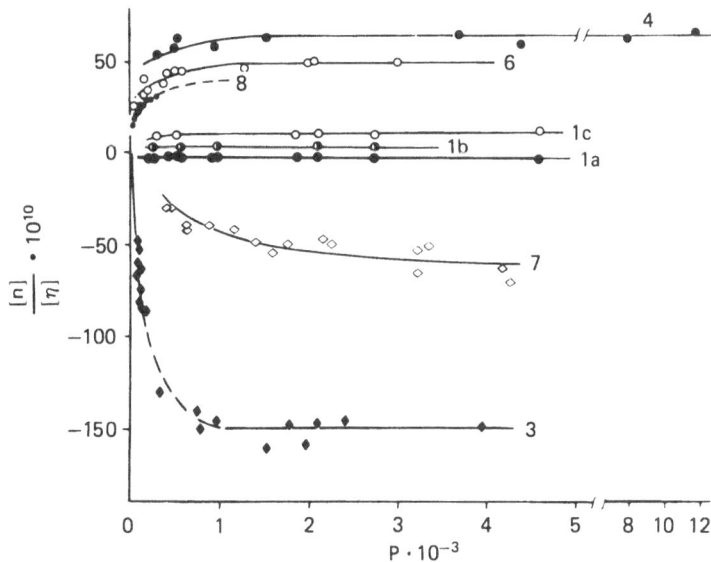

Fig. 21. $[n]/[\eta]$ vs. P (degree of polymerization) plot for cellulose ether and ester solutions. Numbers on curves correspond to those in Table 6

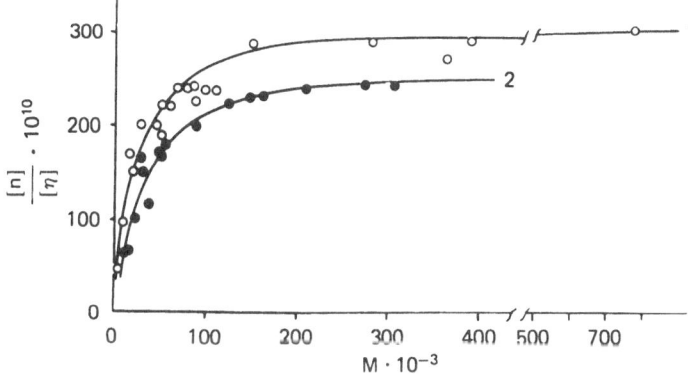

Fig. 22. $[n]/[\eta]$ vs. M plot for of poly(alkyl isocyanate) solutions. 1: Poly(butyl isocyanate) in tetrachloromethane[137]; 2: poly(chlorohexyl isocyanate) in tetrachloromethane[138]

strated by the fact that over the range of molecular weights considered the conformational properties of the molecules of these polymers differ from those of Gaussian chains and approach those of a straight rod, i.e. the molecule behave like "semi-rigid" chains.

For a quantitative comparison of experimental curves in Figs. 20–22 with the theoretical relationship discussed in Chap. 3 experimental data should be represented as a generalized dependence of the ratio $\Delta \equiv ([n]/[\eta])/([n]/[\eta])_\infty$ on $x = 2$ L/A where $([n]/[\eta])_\infty$ is an asymptotic limit of the curves in Figs. 20–22. The values of x for each polymer investigated are given by $x = 2$ M $\lambda/M_0 A$ (Eq. (6) p. 104) in which the length of the Kuhn segment A is known from hydrodynamic data (Tables 1–3). The experimental values of $([n]/[\eta])/([n]/[\eta])_\infty$ obtained in this manner are denoted by points in Fig. 23.

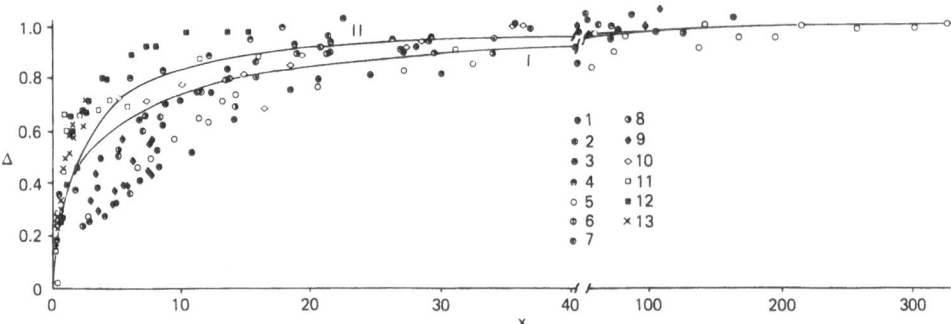

Fig. 23. $([\eta]/[\eta])/([\eta]/[\eta])_\infty = \Delta$ vs. x plot. Points denote experimental data for various poly-mers. 1–3: Ladder polyphenylsiloxane in bromoform[28, 30, 32, 33]; 4: ladder poly(3-phenyl-1-butenesiloxane) in butyl acetate[31, 33, 34]; 5: ladder poly(phenyl-isobutylsiloxane) (1:1) in benzene[33]; 6: ladder poly(phenyl-isohexylsiloxane) (1:1) in benzene[33]; 7: ladder poly(*m*-chlorophenylsiloxane) in tetrachloromethane[33]; 8: ladder polydichlorophenylsiloxane in tetra-bromoethane[38]; 9: cellulose nitrate in cyclohexanone[120]; 10: cellulose carbanilate in diox-ane[116]; 11: poly(butyl isocyanate) in tetrachloromethane[137]; 12: poly(chlorohexyl isocyanate) in tetrachloromethane[138]; 13: poly(γ-benzyl-L-glutamate) in dichloroethane[139]. Theoretical curves: I = according to Noda-Hearst in Ref.[91]; II = Curve 4 in Fig. 18

Experimental points for various polymers are scattered over a wide range in Fig. 23 and cannot be approximated by a single curve. This means that the dependence of Δ on the chain length for all the polymers studied cannot be expressed as a universal dependence of Δ on x containing a single parameter: persistent length a (or the length of the Kuhn segment A) of an equivalent worm-like model.

However, some regularities can be observed in this cloud of points. Thus, the experimental curves (that could be plotted through the points for each polymer) vary for different polymers in the initial slope at low x and in the speed at which they approach the limiting value at high x. A comparison of this behavior of the curves with the data in Tables 1–3 for the corresponding polymers reveals that for polymers with the highest equilibrium chain rigidity (poly(γ-benzyl-L-glutamate), PBLG; poly(alkyl isocyanate)s, PAIC), the slopes of curves $\Delta = \Delta (x)$ are much higher than for ladder polymers with moderate chain rigidity. This is particularly manifested at low x values where the experimental points for most samples investigated are located much lower and for PBLG and PAIC much higher than those fitting to theoretical Curves I and II. These curves describe flow birefringence in terms of a model of an infinitely thin worm-like chain with the shape asymmetry of a kinetic unit p = ∞.

4.1.2 Influence of Chain Diameter Finiteness

The negative deviations of the experimental initial slopes from the theoretical depen-dences I or II for polymer molecules with low equilibrium rigidity can be understood qualitatively by taking into account the finite character of d, the diameter of a real polymer chain, which was not included in the theories discussed in Chap. 3.

Actually, according to Eq. (32), at the limit of chain length (x → 0) the asymmetry of chain shape increases infinitely (p⟶∞). In theoretical Curves I and II in Fig. 23, p = ∞ for all values of x. In fact, for any real polymer chain of length L and diameter d, the shape asymmetry cannot be greater than the value of L/d. This has been considered in Refs.[82, 112] where the calculation of the shape asymmetry of a chain molecule at low x according to Eq. (63) is proposed:

$$p_x = (A/d) \left[(x - 1 + e^{-x})/2\right]^{\frac{1}{2}} \tag{63}$$

In practice this means that in addition to factor b an analogous coefficient b_0 should be introduced into the right-hand side of Eq. (59) (or of any other equation describing the value of $[n]/[\eta]$)

$$b_0 \equiv (p_x^2 - 1)/(p_x^2 + 1) = \frac{x - 1 + e^{-x} - 2\,(d/A)^2}{x - 1 + e^{-x} + 2\,(d/A)^2} \tag{64}$$

When the chain length (i.e. x) increases, factor b_0 rapidly becomes unity. Therefore, if the chain thickness d both at moderate and high x is taken into account this does not introduce any changes into the birefringence theory.

Conversely, at low x when the chain length decreases, b_0 decreases to zero and changes its sign having the limiting value $b_{0,x \rightarrow 0} = -1$. This means that flow birefringence should also change its sign which is confirmed by the experimental data obtained in the study of the anisotropy of chain molecules with long-chain side groups (for which d/A increases with growing length of the side groups). In these cases, the sign of flow birefringence always changes from negative to positive when passing from a polymer chain to a monomer[6, 113]. Similar phenomena are observed in the investigation of the dynamo-optical properties of graft copolymer solutions[114,115].

For polymer molecules with high equilibrium chain rigidity the d/A ratio is low (< 0.1) and b_0 is virtually equal to unity over the whole range of x values available experimentally.

In contrast, for flexible-chain polymers for which d/A is relatively high, according to Eq. (64) at low x the value of b_0 can differ greatly from unity and, correspondingly, the ratio $[n]/[\eta]$ can be lower than the theoretical value for an infinitely thin chain. This is illustrated in Fig. 24 where Curves 1–5 depict the theoretical dependence of $\Delta \equiv ([n]/[\eta])/([n]/[\eta])_\infty$ on log x taking into account the finite value of d at various values of the parameter d/A in Eq. (64).

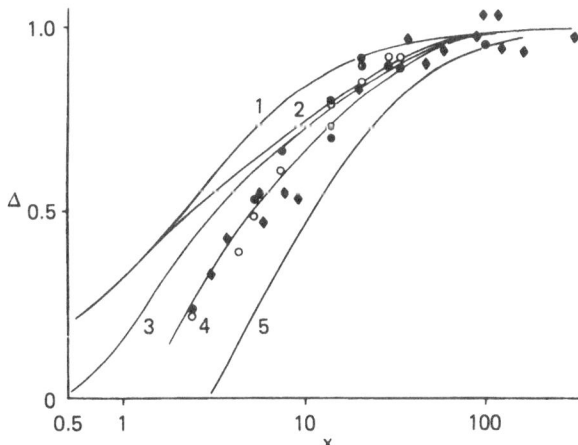

Fig. 24. $\Delta = ([n]/[\eta])/([n]/[\eta])_\infty$ vs. x plot at various A/d parameter values. Curve 1: according to Eq. (59) at $p_x = \infty$ and A/d = ∞; curves 2–5: according to the Noda-Hearst theory[91] with additional factor b_0 according to Eq. (64) at d/A = 0 (2); 0.25 (3); 0.5 (4); 1 (5). Points designate experimental data: cellulose carbanilate in dioxane[116] ($\blacklozenge$); ladder polydichlorophenylsiloxane in bromoform (○) and in tetrabromoethane (●)[38]

The points in Fig. 24 represent the experimental values of Δ for a ladder poly-dichlorophenylsiloxane[38] and cellulose carbanilate[116]. For both polymers the experimental data are in agreement with theoretical Curve 4 corresponding to the value of $d/A = 0.5$ for a kinetically flexible chain polymer. This qualitatively demonstrates that the hydrodynamic properties of the molecules of these two polymers at low x differ from those of an infinitely thin worm-like model. However, to obtain quantitative agreement between theory and experimental data according to Curve 4 in Fig. 24, A/d should be equal to 2 – a reasonable value for many flexible-chain polymers not realistic for such rigid-chain polymers as ladder polysiloxanes or cellulose ethers and esters.

It is interesting that experimental data for polymers with high equilibrium rigidity (PBLG and poly(alkyl isocyanate)) are in good agreement with theoretical Curve III in Fig. 23 which describes the dependence Δ (x) for a kinetically rigid worm-like chain (Eq. (59)) with the limiting value of the shape asymmetry $p_\infty = 2.5$ (similar to Curve 1 in Fig. 18). This implies that polymers with high equilibrium rigidity are also characterized by high kinetic rigidity that should be taken into account in the phenomenon of flow birefringence in accordance with the theory leading to Eq. (59). Hence, for polymers of this type the experimental study of the dependence of $[n]/[\eta]$ on chain length (molecular weight) and the comparison with the theoretical dependence in Eq. (59) can serve as an independent method of determining the parameters of the equilibrium rigidity A and the optical anisotropy β of the chain.

Unfortunately, this procedure is not possible for various classes of polymers with moderate equilibrium chain rigidity shown in Fig. 23 since for all of them the dependence of $[n]/[\eta]$ on M cannot be expressed by a universal function $\Delta = \Delta$ (x). The fact that for these polymers the experimental dependences Δ (x) differ greatly from Curve III in Fig. 23 implies that the dynamo-optical properties of their molecules cannot be described in terms of the theory of kinetical rigid chains. Presumably, flow birefringence in solutions of these polymers is related, to a certain extent, to the kinetic flexibility of their chains.

4.2 Characteristic Orientation Angle

The measurement of orientation angles α is experimentally more difficult than that of birefringence, and in some cases a highly sensitive photoelectric method should be used for this purpose[67]. This requirement is particularly important for the determination of characteristic orientation angles $[\chi/g]$ when the measured values require careful extrapolation to both infinitely low shear rates g and zero concentration c. Figure 25 shows as an example experimental data on the dependence of orientation angle α on g for solutions of fractions of a ladder polydichlorophenylsiloxane in tetrachromethane[38].

The initial slopes

$$(\chi/g) = \lim_{g \to 0} \frac{\pi/4 - \alpha}{g}$$

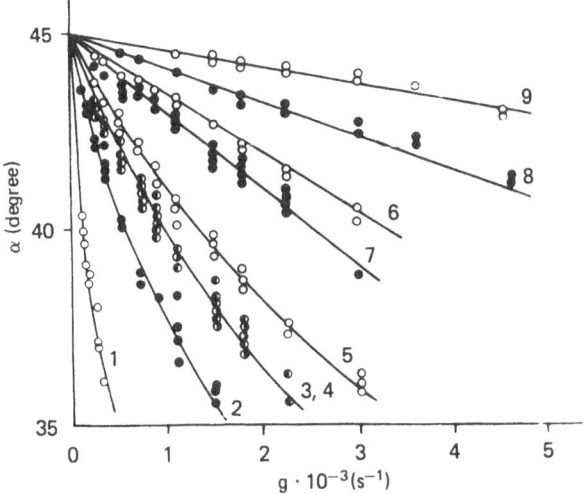

Fig. 25. Orientation angle α vs. velocity gradient g for solutions of ladder polydichlorophenyl-siloxane fractions in tetrabromoethane[38]. Curves correspond to molecular weights M: $2.16 \cdot 10^6$ (1); $0.59 \cdot 10^6$ (2); $0.515 \cdot 10^6$ (3); $0.37 \cdot 10^6$ (4); $0.35 \cdot 10^6$ (5); $0.25 \cdot 10^6$ (7); $0.13 \cdot 10^6$ (8); and $0.09 \cdot 10^6$ (9). Points on each curve designate concentrations the range of which is shown in Fig. 26

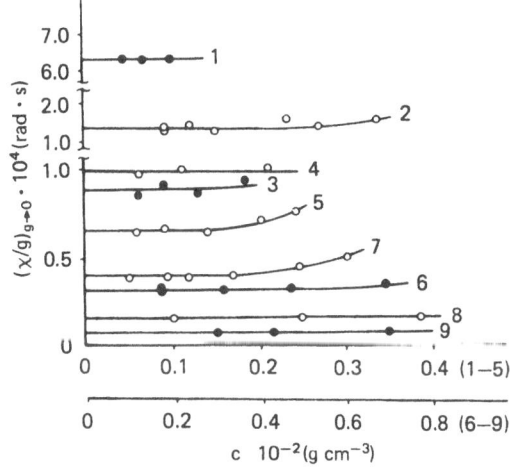

Fig. 26. Initial slopes $(\chi/g)_{g \to 0}$ of orientation angles vs. concentration c for solutions of ladder polydichlorophenylsiloxane fractions in tetrabromoethane[38]. Numbers on curves denote fraction numbers according to Fig. 25

of curves $\alpha = \alpha$ (g) have been determined from their linear parts at α close to $45°$. Figure 26 illustrates the concentration dependence of initial slopes (χ/g) for the same fractions. The extrapolation of the curves in Fig. 26 to $c \longrightarrow 0$ permits a reliable determination of characteristic angles

$$\left[\frac{\chi}{g}\right] = \lim_{c \to 0}\left(\frac{\chi}{g}\right).$$

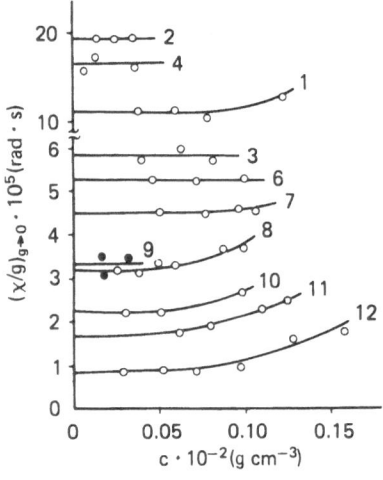

Fig. 27. Initial slopes $(\chi/g)_{g\to 0}$ of orientation angles vs. concentration c for solutions of cellulose carbanilate fractions in dioxane [116]. Molecular weights of fractions measured by sedimentation and diffusion M_{SD} are: 920,000 (1), 870,000 (2), 760,000 (3), 680,000 (4), 440,000 (6), 410,000 (7), 410,000 (8), 280,000 (9), 250,000 (10), 200,000 (11), 190,000 (12)

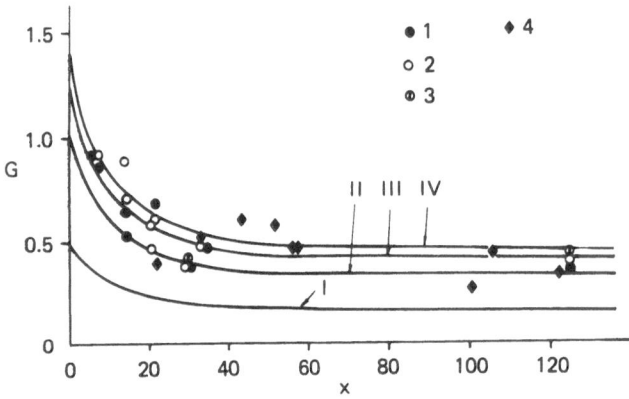

Fig. 28. Plot of G (Eq. (62)) vs. reduced chain length x = 2 L/A for fractions of ladder polydichlorophenylsiloxane in tetrabromoethane (1), bromoform (2), and benzene (3)[38], and cellulose carbanilate in dioxane (4)[116]. Theoretical curves are plotted according to Ref.[91] at the following values of polydispersity parameter U = M_w/M_n: U = 1 (I), 1.4 (II), 1.6 (III), 1.8 (IV)

Similar dependences of (χ/g) on c for solutions of cellulose carbanilate fractions in dioxane[116] are shown in Fig. 27.

These curves also permit the determination of characteristic orientation angles $[\chi/g]$ for the corresponding fractions.

Comparison of the $[\chi/g]$ values with the corresponding values of the $\eta_0[\eta]M$ parameter permits the calculation of coefficient G in Eq. (62) for each sample or fraction.

The dependence of G for fractions of cellulose carbanilate and ladder polydichlorophenylsiloxane on chain length x is described in Fig. 28. Curve I illustrates the Noda-Hearst theoretical dependence[91].

The general character of the experimental dependence is similar to that of the theoretical dependence: the value of G decreases with increasing chain length. However, in terms of absolute magnitude, the experimental values of G greatly exceed the theoretical ones. This is probably mainly due to the molecular weight polydispersity of fractions that was not taken into account in the theory. A high molecular weight polymer is oriented and deformed in flow at lower shear rates than a low molecular weight polymer. Hence, the presence of a high molecular weight fraction should cause an increase in the initial slopes of (χ/g) and, correspondingly, in the G values. Similarly, the polydispersity for "frozen" conformations in an assembly of kinetically rigid chains leads to the same result.

The high sensitivity of orientation angles to molecular weight polydispersity has been widely discussed[68, 69, 130−133]. It has been shown that for a polydisperse sample Eq. (62) holds if coefficient G is replaced by $G_w = \gamma G$ where the correction factor for polydispersity γ is equal to

$$\gamma = \frac{\langle (M^{\alpha+1})^2 \rangle}{\langle M^{\alpha+1} \rangle^2} \cdot \frac{1}{U} \tag{65}$$

$U = M_w/M_n$ is the generally used parameter of polymer polydispersity and α the exponent in the Mark-Kuhn relation (Eq. (10), p. 13). Equation (65) assumes that in Eq. (62) the weight average molecular weight M_w is used as the experimental value of molecular weight.

If the type of molecular weight distribution of the polymer is known, it is possible to express γ as a function of two parameters only, U and α, according to Eq. (65). Thus, for the Schultz-Zimm distribution[134, 135] Eq. (65) yields[133]

$$\gamma = \frac{Z! \, (Z + 1 + 2 \, \alpha)!}{(Z + 1) \, [(Z + \alpha)!]^2} \;,\; Z = \frac{1}{U - 1} \tag{66}$$

For a logarithmic distribution[136] we obtain[133]

$$\gamma = U^{[(\alpha+1)^2 - 1]} \tag{66'}$$

For fractions of a ladder polymer and cellulose carbanilate, α is unity. On the assumption that these fractions are characterized by the Schultz-Zimm distribution the values of γ and the corresponding values of $G_w = \gamma G$ are calculated according to Eq. (66) for three values of U equal to 1.4, 1.6 and 1.8. The Noda-Hearst dependences of G on x corrected in this manner for polydispersity are described in Fig. 28 by Curves II−IV. It is seen that the experimental points agree well with theoretical Curve III for which U = 1.6. This value is close to the values of U obtained in the determination of the molecular inhomogeneity of the same fractions by sedimentation[38]. Hence, the available experimental data on the (χ/g) values and their dependence on molecular weight for polymers with relatively high equilibrium chain rigidity can be interpreted, at least qualitatively, in terms of the existing theories including both the local rigidity of the chain[91] and the polydispersity[133].

4.3 Optical Anisotropy and Structure of Polymer Chains: Ladder Polysiloxanes and Cellulose Esters and Ethers

For kinetically flexible chains of relatively high molecular weight M, when the values of $[n]/[\eta]$ do not vary with M, these values can be used for the determination of the optical anisotropy of a segment $\alpha_1 - \alpha_2 = \beta A$ according to Kuhn's equation[56, 80] equivalent to Eq. (59) if bF_0/F_x is taken to be unity and $x \longrightarrow \infty$

$$[n]/[\eta] = B \, (\alpha_1 - \alpha_2) = B\beta A = BS \, \Delta a \tag{67}$$

where B is determined by Eq. (59').

V. N. Tsvetkov and L. N. Andreeva

Table 5. Limiting values of $[n]/[\eta]$ and anisotropy of the segment $\alpha_1 - \alpha_2$ and the monomer unit $a_\parallel - a_\perp$ of ladder polysiloxanes

Polymer	Solvent	$-\dfrac{[n]}{[\eta]} \times 10^{10}$	$-(\alpha_1 - \alpha_2)$ $\times 10^{25}$ cm^3	$-(a_\parallel - a_\perp)$ $\times 10^{25}$ cm^3	Ref.
1. Ladder polyphenylsiloxane	bromoform	160	1800	25	28, 32, 33)
2. Ladder polyphenylsiloxane	bromoform	110	1230	23	30, 32, 33)
3. Ladder polyphenylsiloxane	bromoform	95	1060	31	30, 32, 33,
4. Ladder polyphenylsiloxane	benzene	240	–	–	36)
5. Ladder poly-*m*-chlorophenyl-siloxane	carbon tetra-chloride	380	4700	40	33, 216)
6. Ladder polydichlorophenyl-siloxane	tetrabromo-ethane	425	4700	53	38)
	bromoform	425	4450	50	38)
7. Ladder poly(3-methyl-1-butene-siloxane)	benzene	47	570	6.5	31)
8. Ladder poly(3-methyl-1-butene-siloxane)	butyl acetate	30	400	4.2	31, 33, 34)
9. Ladder poly(phenyl-isobutyl-siloxane) (1:1)	benzene	69	830	21	33)
10. Ladder poly(phenyl-isohexyl-siloxane) (1:1)	benzene	81	980	19.5	33)
11. Linear poly(methyl-phenyl-siloxane)	benzene	6.1	85.5	17	24, 32)

4.3.1 Ladder Polysiloxanes

In Table 5 are listed the limiting values of $([n]/[\eta])_\infty$ for a number of ladder polysiloxanes according to the data in Fig. 20 and the values of $\alpha_1 - \alpha_2$ calculated by using Eq. (67) and the values of $([n]/[\eta])_\infty$. The last column contains the values of the anisotropy of the monomer unit $\Delta a \equiv a_{||} - a_\perp = (\alpha_1 - \alpha_2)/S$ calculated from $\alpha_1 - \alpha_2$ and S. The number of monomer units in a segment is obtained from Table 2 using the relation $S = A/\lambda$ where λ, the length of the monomer unit in the chain direction, is 2.5 Å according to Fig. 2. For all ladder polymers $([n]/[\eta])_\infty$ is high in absolute values and negative in sign. Hence, the main contribution to the optical anisotropy of the molecule is provided by side groups. This conclusion agrees with the well-known fact that the optical anisotropy of Si–O bonds forming the backbone of the macromolecules is very small[118].

The segmental anisotropy of various ladder polymers is higher by more than one order of magnitude than that of linear poly(methyl-phenylsiloxane). This is in accordance with the fact that the equilibrium rigidity of a ladder structure is more than tenfold greater than that of a single-stranded chain. However, the anisotropy Δa of a monomer unit of ladder polyphenylsiloxanes is only 1.5–2 times greater than that for a linear poly(methyl-phenylsiloxane). This difference is due to the two phenyl rings in the monomer unit of the former and only one ring in the monomer unit of the latter.

This means that the contributions to the anisotropy Δa of a monomer unit provided by one phenyl ring are virtually identical for ladder and single-chain poly-(phenylsiloxane)s. Hence, the double-chain molecular structure does not appreciably affect the rotational mobility of phenyl side groups. The differences in the value of Δa for ladder polymers differing in the structure of side groups are mainly due to varying anisotropies of these groups. Thus, the replacement of a phenyl side group by an alkyl group results in a decrease of the absolute value of Δa whereas the exchange of hydrogen atoms for chlorine atoms in the phenyl ring causes an increase in the negative value of Δa.

Moreover, the differences in the segmental anisotropy of ladder polymers may be related to the differing extent of the defectiveness of their double-chain structure[30, 32, 33] (see p. 100). This can be demonstrated by a comparison of the anisotropy of various samples of ladder polyphenylsiloxanes with an identical chemical structure of monomer units but with the values of $([n]/[\eta])_\infty$ and, hence, of $\alpha_1 - \alpha_2$ which may differ by a factor of 2.5. In fact, if single bonds are introduced into a cyclic chain a *partially* ladder structure results which greatly decreases the equilibrium chain rigidity.

This is true, for example, for poly(N-maleinimide) or for polyacenaphthylene whose chains are only semi-cyclized. In accordance with this their rigidity is much lower than that of a double-chain polymer and only two times higher than that of flexible-chain linear molecules[119]. For ladder polysiloxanes structural defects can appear during polymerization when some hydroxy groups in a polycyclic compound remain unreacted and, correspondingly, the ladder structure in these chain segments remains incomplete. It is easy to show that the differences in the rigidity of various ladder polymers observed experimentally can be caused by very small variations in the degree of defectiveness of their structure. Thus, the occurrence of defects of the ladder structure of the polymer in only 3% of its monomer units is sufficient to decrease by 2.5 times the equilibrium rigidity.

Table 6. Limiting values of $[n]/[\eta]$ and anisotropy of the segment, $\alpha_1 - \alpha_2$ and the monomer unit, $a_{\parallel} - a_{\perp}$ of cellulose ethers and esters

Polymer	Substituting group	Degree of substitution	Solvent	$\dfrac{[n]}{[\eta]} \times 10^{10}$	$(\alpha_1 - \alpha_2) \cdot$ $\times 10^{25}$ cm^3	$(a_{\parallel} - a_{\perp}) \cdot$ $\times 10^{25}$ cm^3	Ref.
1. Cellulose butyrate	$-O-CO-C_3H_7$	3.0	tetrachloroethane	−2.8	−34	−2.6	42)
			bromoform	+3.3	+35	+2.7	
			methyl ethyl ketone	+11.0	+144	+5.6	
2. Cellulose benzoate	$-O-CO-C_6H_5$	3.0	N,N-dimethylformamide	−48.4	−617	−10.4	129)
			chloroform	−60.5	−763	−16.4	
			bromobenzene	−79.3	−914	−19.6	
			dimethylphtalate	−73	−830	−17.8	
3. Cellulose carbanilate	$-O-CO-NH-C_6H_5$	2.2 ± 0.4	dioxane	−150	−1880	−47	116)
		3.0	dioxane 21 °C	−144	−1830	−45.8	125)
			65 °C	−68	−872		
			ethyl acetate	−42	−560	−15.2	
4. Cellulose monophenylacetate	$-O-CO-CH_2C_6H_5$	2.8 ± 0.1	bromobenzene	+52	+600	+13.0	109)
			bromoform	+42	+478	+10.3	
5. Cellulose diphenylacetate	$-O-CO-CH(C_6H_5)_2$		dioxane	+82	+1030		218)
6. Cellulose diphenylphosphonocarbamate	$-O-CO-NH-P(OC_6H_5)_2$ O	2.2 ± 0.3	dioxane	+50	+640	+13.7	109)
7. Cellulose nitrate	$-O-NO_2$	2.75 ± 0.1	cyclohexanone	−65	−824	−14	120)
		1.9	cyclohexanone	−5	−62	−2	83)
			dioxane	+11.7	+149	+4.8	
8. Ethyl cellulose	$-O-C_2H_5$	2.5 ± 0.1	dioxane	+40	+512	+14.6	109)
9. Benzyl cellulose	$-O-CH_2-C_6H_5$	2.5	dioxane	+25	+291		218)
10. 2-Cyanoethyl-acetyl cellulose	$-O-CH_2-CH_2-CN,$ $-O-CO-CH_3$	—	acetone	+29	+390	+17.0	
			N,N-dimethylformamide	+7.5	+90	+3.9	

4.3.2 Cellulose Esters and Ethers

In Table 6 are listed the values of $([n]/[\eta])_\infty$ and the corresponding values of $\alpha_1 - \alpha_2$ and Δa according to Eq. (65) for various cellulose ethers and esters. Δa is calculated from $\alpha_1 - \alpha_2$ and segment lengths A according to Table 1 taking into account the fact that for a polyglucoside chain λ is 5.1 A. Although, according to hydrodynamic data (Table 1), the differences in the equilibrium flexibility of molecules of various cellulose derivatives are not very great, their $([n]/[\eta])_\infty$ values and, hence, those of $\alpha_1 - \alpha_2$ and Δa may vary greatly both numerically and in sign. This means that the optical anisotropy of the chains of cellulose esters and ethers is determined, to a great extent, by the structure and anisotropy of pendant substituents and the degree of substitution (Fig. 29). Thus, the negative anisotropy of the monomer unit greatly increases steeply with the content of anisotropic nitrate side groups.

The microform effect can also substantially affect flow birefringence[67], is induced by a change in the refractive index of the solvent and may influence not only the value but also the sign of the observed anisotropy. According to the data in Table 6 this occurs, for example, for cellulose butyrate and nitrate solutions.

The microform effect can be utilized to evaluate the equilibrium chain rigidity on the basis of dynamo-optical data alone without using hydrodynamic data. Thus the dependence of $[n]/[\eta]$ on the refractive index of solvent for various cellulose ether and ester solutions has been studied (Fig. 30)[121].

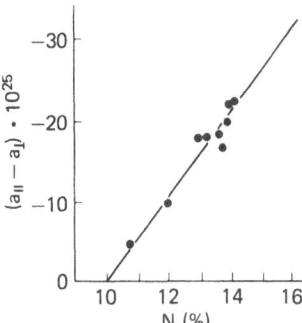

Fig. 29. Graph of anisotropy of cellulose nitrate monomer unit vs. nitrogen content in side groups[120]

Table 7. Number of monomer units S in a segment, segment length A and hindrance parameter σ^2 of cellulose ether and ester solutions according to optical data (microform effect)

Polymer	Degree of substitution	S	A (A)	σ^2
Cellulose nitrate	2.6	50	260	22
Ethyl cellulose	2.6	35	180	15
Cellulose benzoate	3.0	52	270	23
Cellulose carbanilate	1.9	35	180	15
Cellulose monophenyl acetate	2.5	49	250	21
2-Cyanoethyl-acetyl cellulose	1.7	49	250	21
2-Cyanoethyl-trityl cellulose	2.0	54	280	23

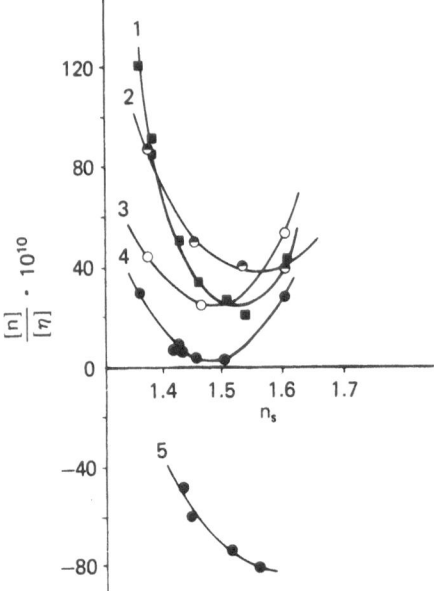

Fig. 30. Plot of $[n]/[\eta]$ of cellulose ether and ester solutions vs. refractive index of solvent[121]: 1: 2-cyanoethyl-tritylcellulose, 2: cellulose monophenylacetate, 3: ethyl cellulose, 4: 2-cyanoethylacetylcellulose and 5: cellulose benzoate

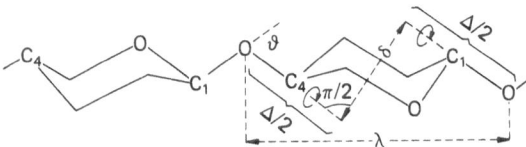

Fig. 31. Conformation of cellulose chain; λ = projection of the monomer unit on chain direction, $\Delta/2 = 2.7$ A = length of an effective bond about which rotation is possible, $\delta = 1.45$ A length of an effective bond about which rotation is impossible

According to Eqs. (67) and (50), the experimental dependence for each polymer has a parabolic shape. Its minimum can be used for the determination of the intrinsic anisotropy of the segment, $S \cdot \Delta a_i$, and the remaining part of the parabola for the calculation of the molecular weight of the segment, $M_0 S$, and the number of monomer units in a segment S according to Eq. (50). The values of S and $A = \lambda \cdot S$ thus obtained are listed in Table 7. They are in satisfactory agreement with the values of A in Table I obtained by sedimentation, diffusion and viscometry.

The experimental data on the equilibrium rigidity of cellulose derivative molecules can be compared with the results obtained by analysis of possible conformations of the polyglucoside chain (Fig. 31).

The flexibility of the chain is caused by a rotation about the $O-C^1$ and $O-C^4$ bonds between neighboring glucose rings. If a real chain is replaced by an equivalent chain each unit of which consists of two parallel $\Delta/2$ bonds about which rotation is possible and one δ bond (normal to the two first bonds) about which no rotation takes place, it can be shown[122] that the number of monomer units S_f in the Kuhn segment of a cellulose chain with unhindered rotation is given by

$$S_f = \left[\left(\frac{\delta}{\Delta} \right)^2 + \frac{1 + \cos \vartheta}{1 - \cos \vartheta} \right] \Bigg/ \left(\cos \frac{\vartheta}{2} + \frac{\delta}{\Delta} \sin \frac{\vartheta}{2} \right)^2 \qquad (68)$$

where $\pi - \vartheta$ is the valence angle between neighboring rotating bonds. Taking $\vartheta = 70°$, $\lambda = 5.4$ A and $\delta = 1.45$ A, we obtain $S_f = 2.3$. A comparison of this value with experimental values yields the degree of hindrance to rotation in a cellulose chain $\sigma^2 = S/S_f$. The values of σ^2 are listed in Table 7. Similar values of σ^2 can be obtained by means A (and hence, S) according to Table 1 determined by hydrodynamic methods.

The hindrance parameter $\sigma^2 \approx 20$ greatly exceeds $\sigma^2 \approx 4$ 5 characteristic of flexible-chain polymers with widely differing structures[10].

This suggests that intramolecular hydrogen bonds formed between side group atoms of a polyglucoside chain play an important role in the hindrance to intramolecular motion of cellulose derivatives.

This suggestion is in agreement with the well-known fact that the equilibrium rigidity of cellulose ether and ester molecules changes greatly with solvent composition and is also confirmed by the strong negative temperature dependence of the statistical size of their chains. The latter property is manifested in a decrease of the intrinsic viscosity, translational friction[124, 125] and the flow birefringence[125, 126] with increasing temperature (Fig. 32)[125].

The backbone rigidity of cellulose derivative molecules containing ionogenic groups can be higher owing to electrostatic interactions between these groups. When the ionic strength of the solution decreases and the polyion chain is correspondingly uncoiled, both $[\eta]$ and the $[n]/[\eta]$ ratio increase. This means that chain uncoiling is accompanied by an increase in the seg-

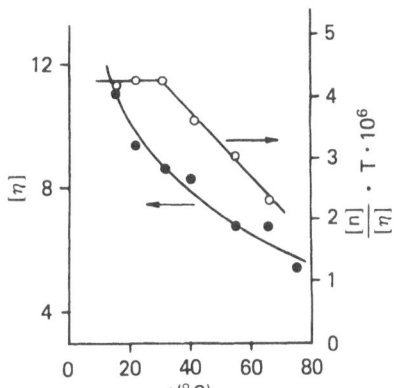

Fig. 32. Temperature dependence of $[\eta]$ and $[n]/[\eta]$ for solutions of cellulose carbanilate in dioxane[125]

Table 8. Hydrodynamic and optical characteristics and equilibrium rigidity of sodium sulfate cellulose molecules at different ionic strengths of the solution I[121, 127]

I (mol/l)	$[\eta]$ (cm^3/g)	$\dfrac{[n]}{[\eta]} \times 10^{10}$ (cm x s^2 g^{-1})	$(\alpha_1 - \alpha_2)$ x 10^{25} cm^3	$\dfrac{\alpha_1 - \alpha_2}{[\eta]}$ x 10^{25} g	A x 10^8 cm
0.200	290	46.6	634	2.19	195
0.150	290	47.4	645	2.22	195
0.100	310	50.0	680	2.19	216
0.010	500	72.0	980	1.96	330
0.005	610	96.7	1320	2.16	410
0.001	870	127	1730	2.00	575

mental anisotropy $\alpha_1 - \alpha_2$ and, correspondingly, the segment length A. These properties are listed in Table 8 which contains the corresponding data for aqueous solutions of a sodium salt of a cellulose sulfoether sample (M = 10^5, pendant group SO_3Na)[127]. It can be clearly seen from these data that the $(\alpha_1 - \alpha_2)/[\eta]$ ratio remains virtually constant as the ionic strength decreases. Hence, since for a Gaussian chain of constant length L the value $\alpha_1 - \alpha_2$ changes proportionally to $\langle h^2 \rangle$ when the coil undergoes polyelectrolytic uncoiling, $[\eta]$ increases proportionally to $\langle h^2 \rangle$, which is characteristic of drained Gaussian coils.

4.4 Polymers with High Chain Rigidity

4.4.1 Dependence of Flow Birefringence on Molecular Weight, Anisotropy and Chain Rigidity

As already mentioned, for polymers with high equilibrium chain rigidity the experimental dependence of $[n]/[\eta]$ on molecular weight agrees with the theoretical relation (59) for kinetically rigid chains. This equation can be utilized for the determination of the parameters A and β from experimental data on flow birefringence of polymer or fraction of varying molecular weight.

For a comparison of the experimental dependence of $[n]/[\eta]$ on M with the theoretical dependence (59) it is convenient to express the parameter x = 2 L/A by M taking into account Eq. (6) (p. 104).

$$x = 2 M \lambda/M_0 A = 2 M/SM_0 = 2 M/M_s \tag{69}$$

where M_s is the molecular weight of the part of the chain with contour length A ("segment molecular weight").

Equation (59) yields the dependence of $[n]/[\eta]$ on M

$$[n]/[\eta] = B\beta A f(2 M/M_s) \tag{70}$$

where $f(2 M/M_s) \equiv y(x)$ is the function given in Table 4.

The initial slope of the curve of the $[n]/[\eta]$ vs. M plot is $[\partial ([n]/[\eta])/\partial M]_{M \to 0} = B \cdot \Delta a/M_0 = B\beta_M$ where β_M is the anisotropy of the extended chain per unit of its molecular weight. The asymptotic limit of the curve is given by $([n]/[\eta])_{M \to \infty} = B\beta_M M_s y_m = B (\alpha_1 - \alpha_2) \cdot y_m$ where the limiting value of y_m is 0.57 (at p_∞ = 2.5). Hence, if experimental data permit the determination of the initial slope of this curve, it is possible to estimate directly β_M or Δa (since M_0 is usually known). The limiting value of the curve determines the anisotropy of the segment, $\alpha_1 - \alpha_2$, and the ratio of the limit to the initial slope equal to $y_m M_s$ allows to ascertain the molecular weight of the segment or its number of monomer units S = M_s/M_0. The same values, β_M and M_s, determine the overall shape of the theoretical curve $[n]/[\eta] = f(M)$.

Curves 1 and 2 in Fig. 22 describe theoretical dependences $[n]/[\eta] = f(M)$ plotted according to Eq. (70) at y_m = 0.57. In this plot such values of the β_M and M_s parameters are selected that ensure the best agreement between theoretical curves and experimental data (points) for poly-(butyl isocyanate)[137] and poly(chlorohexyl isocyanate)[138] in tetrachlormethane. The corresponding values of Δa, β, M_s, S and A are given in Table 9.

Figure 33 shows similar dependences of $[n]/[\eta]$ on M for solutions of poly(γ-benzyl-L-glutamate) (PBLG)[6, 139] (Curve 1). These are compared with data for poly(chlorohexyl isocyanate) (Curve 2). Experimental values of $[n]/[\eta]$ for PBLG do not permit the determination of the limiting value $([n]/[\eta])_\infty$ because they cover only the initial part of the curve $[n]/[\eta] = f(M)$.

In this case a $([n]/[\eta])/M$ vs. M plot is useful since, as it follows from Eq. (70), $([n]/[\eta])/M = [\partial ([n]/[\eta])/\partial M]_{M \to 0} \cdot 2 y(x)/x$. For PBLG this dependence is shown by Curve 3 in Fig. 33. The value of A = 2000 A determined from dynamo-optical properties of this polymer in a helical

Table 9. Parameters characterizing the equilibrium rigidity and optical anisotropy of some rigid-chain polymer molecules according to the data of flow birefringence in their solutions

Polymer	Solvent	M_0	$\left(\dfrac{[n]}{[\eta]}\right)_\infty$ $\times 10^{10}$	β_M $\times 10^{26} \text{cm}^3$	β $\times 10^{17} \text{cm}^3$	Δa $\times 10^{25} \text{cm}^3$	M_s $\times 10^{-3}$	A (Å)	A^* (Å)
Poly(γ-benzyl-L-Glutamate)	dichloroethane	219	–	1.6	17.1	34	220	2000	
Poly(butyl isocyanate)	carbon tetrachloride	99	300	1.47	7.3	14.6	44	890	
Poly(chlorohexyl isocyanate)	carbon tetrachloride	161	250	1.0	8.0	16	54.8	680	
Poly(p-tolyl isocyanate)	carbon tetrachloride	121	–3.1	–0.4	–2.45	–4.9	0.96	16	
Poly-p-benzamide	sulfuric acid	119	–	8.0	14.8	95	16	870	
Poly(p-phenylene terephthalamide)	sulfuric acid	119	400	9.75	18.0	116	9.6	550	620
Poly(amide benzimidazole)	sulfuric acid	119.7	340	8.4	15.0	100	8.9	500	530
Poly(amide hydrazide)	dimethyl sulfoxide	93.7	280	6.8	12.8	63.7	8.8	470	440
Poly(p-phenylene-1,3,4-oxadiazole)	sulfuric acid	114	130	7.9	15.0	114	3.64	190	200
Poly(m-phenylene isophthalamide)	sulfuric acid	119	30	7.3	14.3	87	0.92	47	47

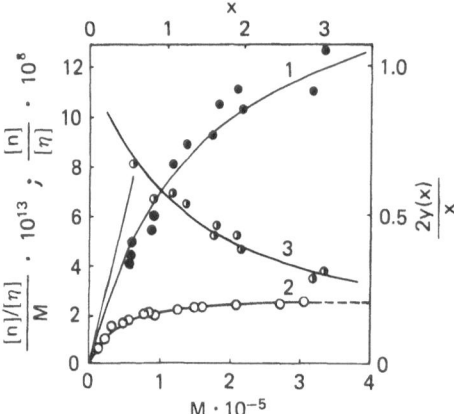

Fig. 33. $[n]/[\eta]$ (1, 2) and $([n]/[\eta])/M$ (3) vs. M plot. 1, 3: poly(γ-benzyl-L-glutamate) in dichloroethane[139], 2: poly(chlorohexyl isocyanate) in tetrachloromethane[138]. The curves 1, 2 describe theoretical dependence according to Eq. (70) at $y_m = 0.57$ and values of molecular parameters are given in Table 9. 3: theoretical dependence of $2y/x$ on x according to Table 4. Points represent experimental data

conformation is in good agreement with the corresponding data obtained from sedimentation and diffusion measurements[140].

Points in Fig. 34 denote experimental values in the $[n]/[\eta]$ vs. M plot for a number of polymers with aromatic chains. For the polymer with the lowest rigidity, poly(m-phenylene isophthalamide), the ratio $[n]/[\eta]$ is virtually constant over the whole range of M investigated and equal to the limiting value $([n]/[\eta])_\infty$ permitting the determination of $\beta A = \alpha_1 - \alpha_2$. The parammeters β and Δa for this polymer are calculated taking into account the known value of A computed from hydrodynamic data (Table 3).

For comparison with poly(alkyl isocyanate)s, Table 9 also provides the optical characteristics of a flexible-chain poly(tolyl isocyanate)[147, 148] which are also calculated from the experimental values of $[n]/[\eta]$ and the value of A from Table 3. The replacement of alkyl groups by aromatic groups destroys conjugation in polyisocyanate chains and causes not only a more than fifty-fold decrease in their rigidity but also a change in the sign of the optical anisotropy of the monomer unit Δa. In poly(tolyl isocyanate) molecules Δa is governed by the anisotropy of the phenyl side group and, hence, is negative, just as in polystyrene[67] and polyphenylsiloxane (Table 5) chains, although its absolute value is much lower.

The experimental values of both $[n]/[\eta]$ and x = 2 L/A differ greatly for various polymers (Table 9 and Fig. 34). However, a single curve can be obtained by plotting $([n]/[\eta])/([n]/[\eta])_\infty$ vs. x (Fig. 35). Curves A and B are theoretical dependences (Eq. (59), p. 126) at $p_\infty = 2.5$, $y_m = 0.57$ and $p_\infty = \infty$, $y_m = 1$, respectively. Curve C corresponds to the Noda-Hearst theory[91]. Experimental data are in good agreement with Curve A showing that for these polymers a kinetically rigid worm-like chain is an adequate model for the dynamooptical properties of their molecules.

For all polymers in Table 9 the ratios $([n]/[\eta])_\infty$ are very high in accordance with the high equilibrium rigidity of their chains. Aromatic polymers are also characterized by a high positive value of the anisotropy of the monomer unit Δa and, correspondingly, by a high positive value of that of unit length β. This is due to the high intrinsic anisotropy of phenyl rings in the main chain and the high refractive index increment dn/dc of the solution-solvent system which drastically increases the microform effect (Eq. (50)).

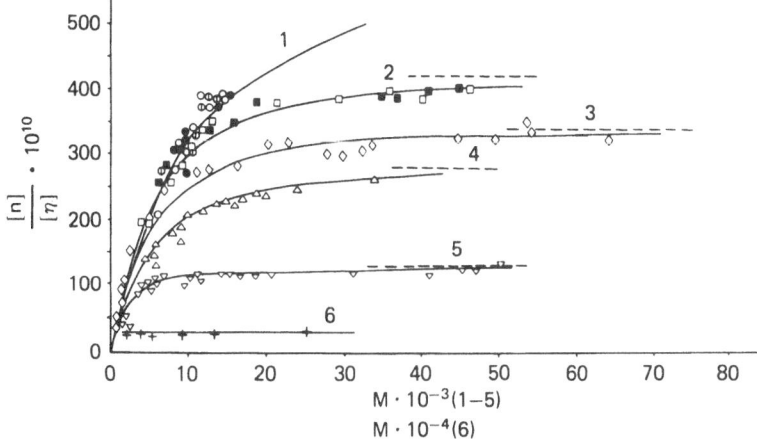

Fig. 34. $[n]/[\eta]$ vs. M plot for some aromatic polymers. Points denote experimental data. 1: poly(p-benzamide) in sulfuric acid. Values of $[n]/[\eta]$ are in accordance with Ref.[49] and molecular weights with Ref.[54] (●), Ref.[55] (Φ), and Ref.[141] (○), respectively, 2: poly(p-phenylene terephthalamide) in sulfuric acid. Values of $[n]/[\eta]$ as reported in Ref.[142] and molecular weights as reported in Ref.[143] (■) and Ref.[141] (□), 3: poly(amide benzimidazole) in sulfuric acid[144], 4: poly(amide hydrazide) in dimethyl sulfoxide. Values of $[n]/[\eta]$ are in accordance with Ref.[145] and molecular weights with Ref.[46], 5: poly(p-phenylene-1,3,4-oxadiazole) in sulfuric acid[47], 6: poly(m-phenylene isophthalamide) in sulfuric acid, values of $[n]/[\eta]$ according to Ref.[146] and molecular weights according to Ref.[45]. Solid curves depict the theoretical dependences calculated according to Eq. (70) by using the structural chain parameters listed in Table 9

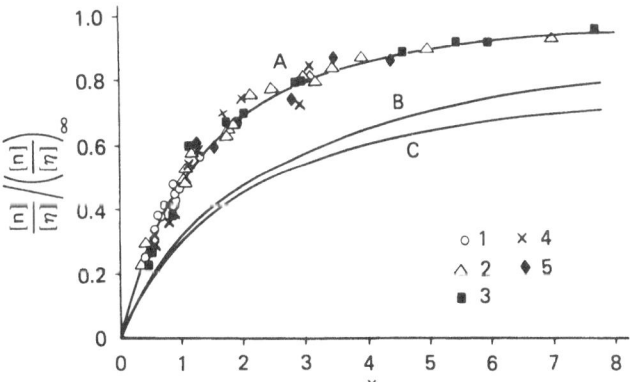

Fig. 35. Graph of relative values of $([n]/[\eta])/([n]/[\eta])_\infty$ vs. parameter x for poly(p-benzamide) (1), poly(amide hydrazide) (2), poly(chlorohexyl isocyanate) (3), poly(γ-benzyl-L-glutamate) (4), and poly(amide benzimidazole) (5)

4.4.2 Orientation Angle and Effect of Polydispersity

High birefringence $\Delta n/g$ in solutions of aromatic polymers facilitates measurements of orientation angles α. Moreover, since the molecular weights of these polymers are usually low, the measured values of α differ only by several degrees from $\pi/4$ and the dependence of α on g in this

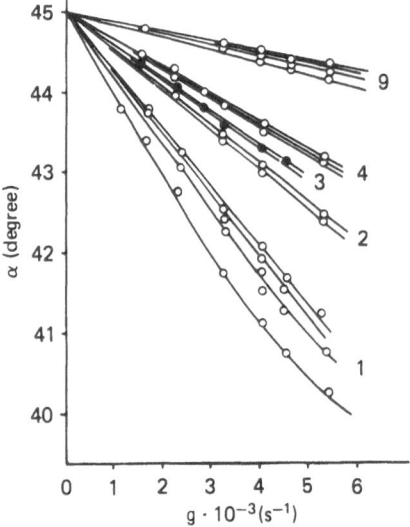

Fig. 36. Orientation angle α as a function of velocity gradient g of samples 1–4 and 9 of poly(amide hydrazide) in dimethyl sulfoxide differing in molecular weight. Several curves for one sample correspond to different concentrations the range of which is shown in Fig. 37 [145]. The curves correspond to molecular weights M: 34,000 (1), 24,000 (2), 19,000 (3), 20,000 (4), and 12,000 (9) [46]

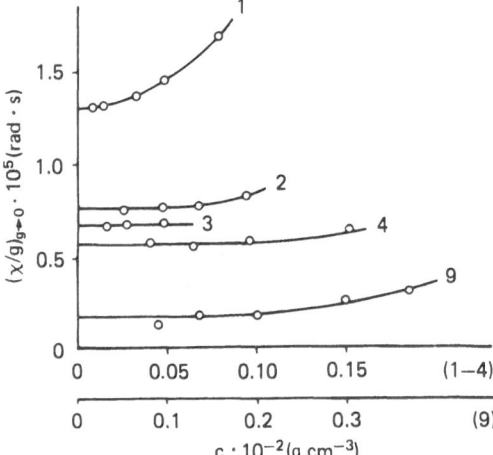

Fig. 37. $(\chi/g)_{g \to 0}$ vs. c plot for various samples of poly(amide hydrazide) in dimethyl sulfoxide [145]. Numbers on curves denote sample numbers in Fig. 36

range is essentially linear. This is shown in Fig. 36 where the corresponding dependences are plotted for several poly(amide hydrazide) samples of different molecular weights. The straight lines in Fig. 36 permit a reliable determination of initial slopes $(\chi/g)_{g \to 0}$ and the use of relatively low polymer concentrations c (since Δn is high) allows a secure extrapolation of these slopes to c → 0 (Figs. 37 and 38).

According to Eq. (62), the characteristic orientation angles [χ/g] thus obtained can be compared with intrinsic viscosities [η] and molecular weights M if the latter are known. In this case coefficient G in Eq. (62) (p. 128) may be determined according to the graph in Fig. 39. The values of M used for this purpose have been ascertained from light scattering, diffusion and viscometric measurements. Moreover, the molecular weights of segments, M_s, given in Table 9 were used for the determination of the corresponding values of x.

These G values are slightly higher than the theoretical values for kinetically rigid drained coils (denoted on the ordinate by B) and increase with decreasing chain length for all the polymers investigated. This may be due to some polydispersity of samples the effect of which should probably increase with decreasing characteristic angles [χ/g] [47].

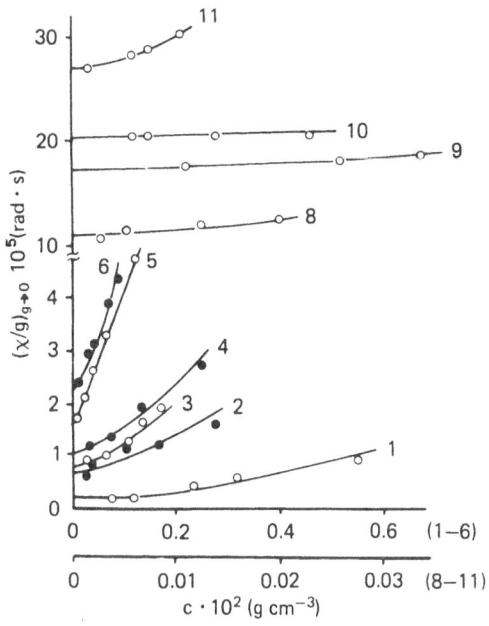

5

Fig. 38. $(\chi/g)_{g \to 0}$ vs. c plot of poly(p-phenylene terephthalamide) samples in sulfuric acid. The curves correspond to following molecular weights M [143]: 4500 (1), 6500 (2), 7300 (3), 8900 (4), 12,500 (5), 16,000 (6), 35,000 (8), 41,000 (9), 42,000 (10), 45,000 (11)

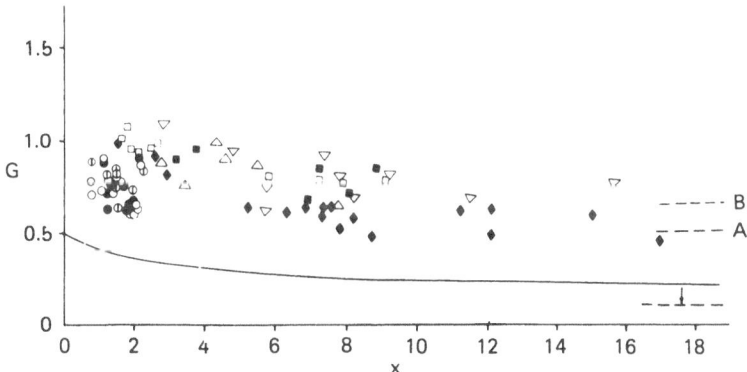

Fig. 39. G vs. x plot for various aromatic polymers samples: 1: poly(p-benzamide) in sulfuric acid. Values of [χ/g] according to Ref. [49]. Molecular weights are plotted according to data in Refs. [54, 55, 142] (⊘, • and ○ respectively); 2: poly(p-phenylene terephthalamide) in sulfuric acid. Values of [χ/g] plotted according to data in Ref. [142] and molecular weights according to data in Refs. [143, 141] (■ and □ respectively); 3: poly(amide hydrazide) in dimethyl sulfoxide. Values of [χ/g] plotted according to data in Ref. [145] and molecular weights according to data in Ref. [46] (△); 4: poly(amide benzimidazole) in sulfuric acid [144] (♦); 5: poly(p-phenylene-1,3,4-oxadiazole) in sulfuric acid [47] (▽). The curve corresponds to theory as in Ref. [91]. A and B are theoretical values for kinetically rigid Gaussian coils with strong (A) and weak (B) hydrodynamic interactions [93]

However, the fact that for this class of polymers the experimental values of G exceed the theoretical one $G = 0.67$[93] by about 30% in the low molecular weight range of x is of great practical importance. This fact supports the use of Eq. (62) (p. 128) for the determination of the molecular weight M of the polymer from experimental values of $[x/g]$, $[\eta]$ and η_0 and the theoretical value of G[47, 49, 142, 145, 146]. This method ensures a ready determination of the molecular weight of many rigid-chain aromatic polymers whereas the application of other methods for the determination of M is connected with experimental difficulties (see Sect. 2.2).

It should be borne in mind that characteristic orientation angles are very sensitive to molecular weight polydispersity of the polymer, in particular, to the presence of high molecular fractions. Hence, for each real polymer sample the molecular weight $M_{x\eta}$ ascertained from $|x/g|$ and $[\eta]$ using Eq. (62) and the theoretical value of G should exceed both the weight average molecular weight M_w and the average weight $M_{D\eta}$ obtained from the experimental values of $[\eta]$ and D by use Eq. (11) and the theoretical value of the A_0 parameter. Thus, it may be concluded that the polydispersity of aromatic polymers shown in Fig. 39 is relatively low since their experimental G values only exceed slightly the theoretical limit for a monodisperse polymer.

Molecular weight polydispersity may also markedly affect the determination of the equilibrium rigidity of the macromolecule from the experimental dependence of $[n]/[\eta]$ on M. In fact, M_s (and hence A) is governed by the ratio of the limiting value of $([n]/[\eta])_\infty$ relative to the initial slope of the curve $[n]/[\eta] = f(M)$. $([n]/[\eta])_\infty$ is independent of molecular weight and therefore insensitive to polydispersity. In contrast, in the initial range of the curve ($x \rightarrow 0$), $[n]/[\eta]$ increases linearly with M and for a polydisperse sample it depends on the ratio $\langle W(\gamma_1 - \gamma_2)/M \rangle / \langle W/M \rangle$. For a rod-like molecule this ratio is proportional to $\langle M^3 \rangle / \langle M^2 \rangle = M_z$. Since M_w and $M_{D\eta}$ are, generally, used, the plot of $[n]/[\eta]$ vs. M may lead to overestimated values of the initial slope and, too high values of Δa and consequently to too low values of M_s and A.

The fact that $([n]/[\eta])_\infty$ is independent of the molecular weight of the polymer prevents an error in the determination of S or A caused by the polydispersity of the sample. For this purpose, it is sufficient to use the value of $([n]/[\eta])_\infty$ for a given polymer comparing it not with the initial slope of curve $[n]/[\eta] = f(M)$, but with the value of $([n]/[\eta])_\infty$ of another polymer for which the length of the Kuhn segment A is known (e.g., from hydrodynamic data). Then the anisotropy of unit length β is close to that of the polymer investigated. The absolute values of β are required.

The comparison is based on Eq. (70) according to which for two polymers 1 and 2

$$([n]/[\eta])_{1,\infty}/([n]/[\eta])_{2,\infty} = \beta_1 A_1/\beta_2 A_2 = \Delta a_1 S_1/\Delta a_2 S_2 \qquad (71)$$

Applying this method to aromatic polymers listed in Table 9, assuming that $\beta_1 = \beta_2 = \beta_3 = \ldots$, and using as a standard for comparison poly(m-phenylene isophthalamide) for which A = 47 A (from hydrodynamic data in Table 3) the values of A* given in the last column of Table 9 can be obtained. These values are in reasonable agreement with the values of A in the same Table.

In principle, it is possible to determine the absolute value of the anisotropy $\beta_M = \Delta a/M_0$ of the polymer chain by comparing the experimental values of $[n]$ and $[x/g]$ for a polymer solution. From Eqs. (58) and (61) we readily obtain

$$[n]/[x/g] = B_0 \beta_M \Psi(x) \qquad (72)$$

where $B_0 = 8 \pi N_A (n^2 + 2)^2/45 n \eta_0$.

The function

$$\Psi(x) = b \frac{\langle h^4 \rangle}{\langle h^2 \rangle} \cdot \frac{\langle \gamma_1 - \gamma_2 \rangle}{\beta L} \qquad (73)$$

can be tabulated[61] for using Eqs. (32) (p. 118), (47) (p. 123), (48) (p. 123), (52) (p. 125) and (57) (p. 126).

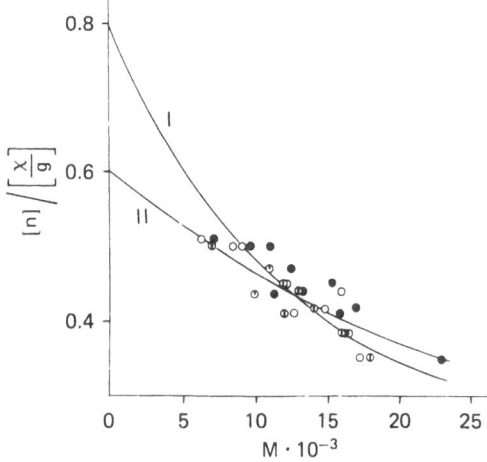

Fig. 40. Ratio of characteristic values of $[n]$ to $[\chi/g]$ vs. molecular weight M for poly(p-benzamide) in sulfuric acid. Values of $[n]$ and $[\chi/g]$ are plotted according to data in Ref.[49] and molecular weights M_W (⊕), M_W (○) and M_η (●) according to Refs.[55, 141], respectively. The curves illustrate theoretical dependences plotted according to Eq. (72) for the following parameters: I. $\Delta a = 55 \cdot 10^{-25}$ cm^3, A = 970 A; II. $\Delta a = 42 \cdot 10^{-25}$ cm^3, A = 2000 A

Function $\Psi(x)$ depends on the polydispersity of the polymer at all values of x except the limiting case $x \longrightarrow 0$ when $\Psi(x) \longrightarrow 1$ regardless of the molecular weight distribution in an assembly of thin rod-like molecules. Hence, the determination of the ratio $[n]/[\chi/g]$ for a polymer homologous series and extrapolation to $M \longrightarrow 0$ permits to ascertain $\beta_M = \Delta a/M_0$ according to Eq. (72) irrespective of the polydispersity of the polymer.

However, this method requires high precision in the determination of the experimental values of $[\chi/g]$ at low M and, correspondingly, at low angles χ when the chain conformation is close to rod-like (Fig. 40). The scattering of experimental points and the absence of experimental data on orientation angles at low M leads to a relatively arbitrary selection of extrapolated curves and thus to a wide variation in possible values of Δa and A. This method may become efficient after a further development of the measuring technique of the orientation angle.

4.5 Optical Anisotropy and Structure of Polymer Chains: Aromatic Polymers

Data on the optical anisotropy and equilibrium rigidity of aromatic polymers can be used to draw conclusions on their structural and conformational properties.

4.5.1 Aromatic Polyamides

This possibility is particularly useful for the characterization of aromatic polyamides[101] which are of great practical importance in the manufacture of synthetic materials.

As already mentioned, the very high chain rigidity of aliphatic polyamides and poly(alkyl isocyanate)s is due to the energy of quasiconjugation in amide groups located at a minimum distance apart along the chain.

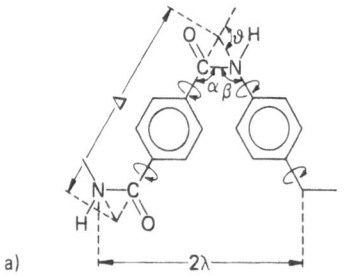

a)

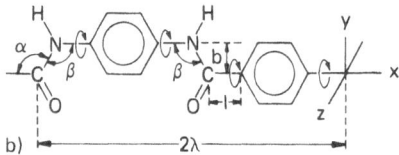

b)

Fig. 41 a, b. Conformation of the poly(p-phe-nylene terephthalamide) chain exhibiting *cis*-(a) and *trans*-(b) structure of the amide group

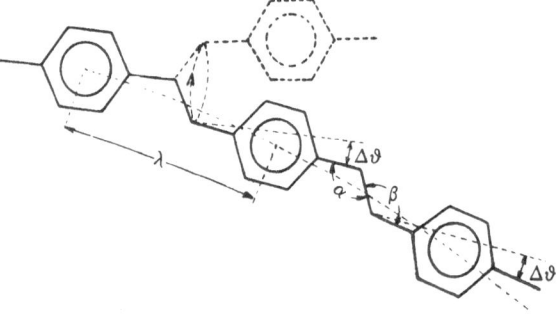

Fig. 42. Conformation of the *para*-aromatic polyamide chain with different angle α and β at the carbon and nitrogen atom of the amide group

In aromatic polyamides the amide groups are separated along the chain by phenyl rings introducing into it bonds about which rotation is possible. Hence, the conformation and equilibrium rigidity of the chain greatly depend on whether the rings are included in the chain in the *meta*- or *para*-position or whether the amide group adopts *cis*- or *trans*-conformation.

The simplest *para* aromatic polyamide, poly(p-phenylene terephthalamide) (PPPhTPhA), is an example showing that the presence of a considerable amount of amide groups in *cis*-configuration (Fig. 41 a) is impossible because in this case chain flexibility would have been many times lower than the experimental value $A \approx 600 \text{ Å}$ (Table 9). In contrast, if the amide group adopts *trans*-configuration (Fig. 41b), the chain of *para*-aromatic polyamide exhibits a "crankshaft"[106] conformation in which all axes of internal rotation in the chain are parallel to each other. The molecule then acquires a "rod-like" shape which ensures very high equilibrium rigidity found experimentally. The fact that the experimental value of A is finite although it is very high can be quantitatively explained by small differences in the valence angle at the carbon and nitrogen atom of the amide groups leading to chain "bending" (Fig. 42).

Moreover, this "bending" and possible differences in the hindrance to intramolecular rotations explain[149-151] the variation in the segment length A between poly-(p-benzamide) and poly(p-phenylene terephthalamide) molecules (Table 9).

It should be taken into account that in very rigid chains, such as those of poly-(alkyl isocyanate)s and *para*-aromatic polyamides, apart from rotation about valence bonds another mechanism can contribute to flexibility: the deformation of valence angles and bonds during thermal chain motion just as it should occur in ladder structures (p. 100). When several flexibility mechanisms exist, the resulting rigidity of the homopolymer chain can be evaluated if the flexibilities, resulting from different mechanisms, are considered to be additive and the following equations are used:

$$\frac{1}{S} = \sum_i \frac{1}{S_i} \, , \, \frac{1}{A} = \sum_i \frac{1}{A_i} \tag{74}$$

where A_i is the length of the Kuhn segment and S_i the number of monomer units in this segment when only one i-th mechanism in the chain exists and A and S correspond to all simultaneously occuring mechanisms[106].

When the aromatic rings are incorporated into the chain of an aromatic polyamide in the *meta* position, every ring introduces into the chain a rotating bond forming with the two neighboring bonds an angle of 60°. The equilibrium rigidity steeply decreases although the amide groups adopt a *trans*-configuration (see Fig. 43 and Table 9 for poly(*m*-phenylene isophthalamide (PMPhIPhA)).

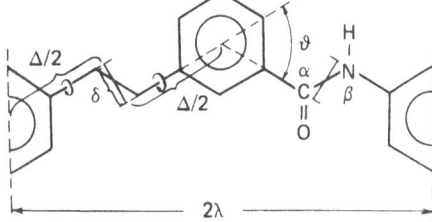

Fig. 43. Conformation of an extended chain of poly(*m*-phenylene isophthalamide) with *trans* structure of the amide group. Angles α and β are assumed to be equal, λ is the projection of the monomer unit on the chain direction

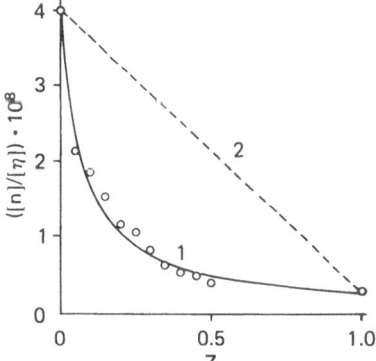

Fig. 44. [n]/[η] vs. Z plot (Z = relative content of *meta*-phenyl rings in the chain) for solutions of some aromatic polyamide in copolymers in sulfuric acid (points indicate experimental data). Curves 1, 2: see text

4.5.2 Aromatic Copolymers

The decrease in the equilibrium rigidity of a *para*-aromatic polyamide molecule through incorporation of aromatic rings into the *meta*-position in the chain is also shown in Fig. 44. The points denote the experimental values of [n]/[η] for some high molecular weight copolymers

Table 10. Shear optical coefficient $\Delta n/\Delta\tau$ and equilibrium rigidity of some amide- and ring-containing polymers in solutions

Polymer	Monomer unit	Solvent	$(\Delta n/\Delta\tau)$ $\times 10^{10}$ $(\mathrm{g}^{-1}\,\mathrm{cm}\,\mathrm{s}^2)$	A (Å)
1. Poly(N,N'-di-methyl-p-phenyl-ene terephthal-amide)		sulfuric acid	20	27
2. Heterocyclic polyamide		sulfuric acid	340	500
3. Poly(amide acid)		N,N-dimethyl-acetamide	70	90
4. Poly(amidoimide)		sulfuric acid	270	400
5. Poly(amide acid)		N,N-dimethyl-acetamide	60	90

			200–350	300–500
6. Poly(pyromellite imide of benzidine)		sulfuric acid	200–350	300–500
7. Polycaprolactam	$\cdots -(CH_2)_5-CO-NH-\cdots$	sulfuric acid	5	18
8. Poly(cyclohexane amide)		sulfuric acid	32	150

of *para* and *meta* aromatic polyamides[154] differing in the ratios Z of the phenyl rings in the *meta* (m_1) and *para* (m_2) positions: $Z = m_1/(m_1 + m_2)$. The ratio $[n]/[\eta]$, which is proportional to the length of the Kuhn segment A, decreases sharply with increasing Z (at low Z) and approaches $[n]/[\eta]$ for PMPhIPhA (at Z = 1). These data reveal the validity of the additivity principle for the mechanisms of flexibility expressed by Eqs. (74).

For the copolymers investigated these equations are equivalent to

$$1/([n]/[\eta]) = Z/([n]/[\eta])_1 + (1 - Z)/([n]/[\eta])_2 \tag{75}$$

where $([n]/[\eta])_1$, $([n]/[\eta])_2$ and $[n]/[\eta]$ refer to poly(*m*-phenylene isophthalamide) (PMPhIPhA) (Z = 1), poly(*p*-phenylene terephthalamide) (PPPhTPhA) (Z = 0) and their copolymer (Z), respectively.

Full Curve 1 in Fig. 44 shows the values of $[n]/[\eta]$ calculated according to Eq. (75) from the given values of $([n]/[\eta])_1$, $([n]/[\eta])_2$ and Z. Curve 1 is in satisfactory agreement with the experimental points in contrast to broken line 2 which corresponds to the additivity of the rigidities of the components in the copolymer. This curve is expressed by

$$S = ZS_1 + (1 - Z) \cdot S_2 \tag{76}$$

Hence, the tremendously high equilibrium rigidity and the ordered structure of *para* aromatic polyamides favouring the formation of mesophases in the concentrated polymer solutions and permitting their use for the manufacture of ultrahigh-modulus fibers[107, 108], is ensured by both the *trans*-structure of the amide groups and the *para* position of the phenyl rings.

If one of these structural conditions is not fulfilled, the molecule loses its unique properties. This is exemplified by *N*-methyl substituted poly(*p*-phenylene terephthalamide)[152]. Its equilibrium rigidity (Table 10) is 20 times lower than that of the unsubstituted analog (Table 9) and approaches that of poly(*m*-phenylene isophthalamide). Probably, in this case, the coplanar *trans*-structure of the amide group is distorted by steric interactions of the pendant methyl group as in poly(alkyl isocyanate)s.

The fact that the rigidity of poly(amide hydrazide) molecules is lower than that of PPPhTPhA can also be explained by the presence of "defective" amide groups (with *cis*-structure) in the former[145] (Table 9).

The structures of the chain of poly(amide benzimidazole) (Tables 9 and 10) and of aromatic polyamides differ in the presence of the heterocycle

R = NH or O

in the repeating unit of the former. When this heterocycle is incorporated into the chain, a rotating bond at an angle of 20–30° is introduced into the chain and this slightly lowers the chain rigidity as compared to that of *para* aromatic polyamides.

The formation of uniform rings in the chain owing to covalent (ladder polymers) or hydrogen (cellulose derivatives) bonds leads to an increase in its rigidity and, correspondingly, in flow birefringence. A comparison of properties of aromatic polymers with polyimides[153] and of the corresponding poly(amide acids) transformed into polyimides by thermal cyclization offers a good evidence of this effect. The examples given in Table 10 reveal that flow birefringence is very sensitive to imidization: the shear optical coefficient $\Delta n/\Delta \tau$ and the segment length A increase by more than an order of magnitude when the polyacid is transformed into the corresponding polyimide.

4.5.3 Hindrance to Rotation in Aromatic Chains

The structure of the repeating unit in aromatic polymers can be much more complex than that in common flexible-chain polymers. However, in many cases it is possible to distinguish "virtual" rotating bonds in the aromatic chains and to utilize them for the calculation of the statistical size of the chain assuming free rotation about these bonds.

Thus, according to Fig. 43 for PMPhIPhA each chain unit containing an amide group and a phenyl ring ("monomer unit") may be replaced by two virtual bonds Δ and δ normal to each other. Rotation is possible only about the former, in complete analogy with the cellulose chain unit (Fig. 31). Hence, when rotation in the chain of this polyamide is unhindered, the number of monomer units S_f in its segment can be determined according to Eq. (68) where $\vartheta = 60°$ and $\delta/\Delta = 0.2$[106]. Substitution of these values into Eq. (68) yields $S_f = 3.3$ and, correspondingly, the length of the Kuhn segment

$$A_f = S_f \cdot \lambda = S_f \left(\delta \sin \frac{\vartheta}{2} + \Delta \cos \frac{\vartheta}{2} \right) = 3.3 \times 6.1 \, A = 20 \, A.$$

Comparison of the value of A_f with the experimental value A (Table 3) gives the degree of hindrance to rotation in the PMPhIPhA chain $\sigma = (A/A_f)^{0.5} = 1.5$.

The values of σ for some copolymers of *meta-* and *para*-aromatic polyamides for which the experimental values of $[n]/[\eta]$ are shown in Fig. 44 can be determined similarly. If the simplifying assumption of the uniformity of structure of all repeating units in the copolymer chain is made (Fig. 45), each repeating unit can be replaced by two virtual bonds $\Delta = \Delta_1/Z$ and $\delta = \delta_1/Z$ where Δ_1 and δ_1 are virtual bonds of the PMPhIPhA chain (Fig. 43).

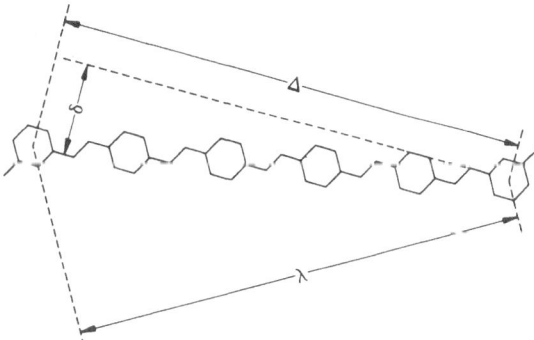

Fig. 45. Model of the repeating chain unit of the copolymer of *meta-* and *para*-aromatic polyamides at $Z = 0.2$

Using this model and taking into account Eqs. (71) and (68) it is possible to relate the shear optical coefficients of the copolymer $\Delta n/\Delta \tau$ and of PMPhIPhA $(\Delta n/\Delta \tau)_1$ to the ratio σ/σ_1 as follows: $(\Delta n/\Delta \tau)/(\Delta n/\Delta \tau)_1 = \{1 + 0.625 \, [(1/Z) - 1]\} \, (\sigma/\sigma_1)^2$ where σ and σ_1 are degrees of hindrance in the chains of the copolymers and PMPhIPhA ($Z = 1$), respectively. This expression permits the determination of σ from the experimental values of $\Delta n/\Delta \tau$ at different $Z \geqslant 0.05$ if σ_1 is known at $Z = 1$. The results presented in Table 11 reveal that for all the copolymers investigated the numerical values of the hindrance parameters are similar.

A comparison of the experimental value of $\Delta n/\Delta \tau$ for poly(tetraphenylmethane terephthalamide) (Fig. 46) and PMPhIPhA solutions also permits the determination of the equilibrium rigidity and degree of hindrance σ [155] (Table 11).

Table 11. Shear optical coefficients $\Delta n/\Delta\tau$, length of the Kuhn segment A and hindrance parameters σ in aromatic polymer chains

Polymer	Z	$(\Delta n/\Delta\tau) \times 10^{10}$ $(g^{-1} \, cm \cdot s^2)$	A (Å)	σ
Copolymers: poly(*meta*-co-*para* aromatic amides)	1.0	26.3	47	1.5
	0.50	37.8	83	1.4
	0.45	47.0	105	1.5
	0.40	52.0	120	1.5
	0.35	61.0	144	1.6
	0.30	81.2	197	1.7
	0.25	105	261	1.8
	0.20	117	296	1.7
	0.15	152	398	1.7
	0.10	186	503	1.6
	0.05	212	587	1.2
Poly(tetraphenyl-methane terephthal-amide)		36	130	1.6
Poly(*p*-phenylene-1,3,4-oxadiazole)		130	200	1.6
Poly(phenyl quino-xaline)	I	140	78	1.2
	II	85	54	1.4
	III	50	34	1.3

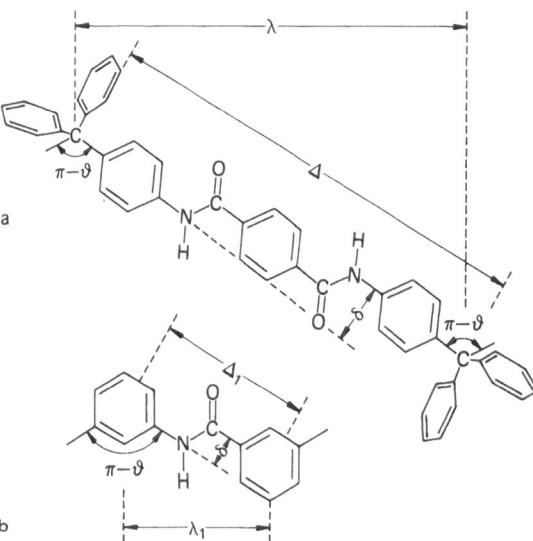

Fig. 46a,b. Structure of the repeating unit of poly(tetraphenylmethane terephthalamide) (a) and poly(*m*-phenylene isophthalamide) (b)

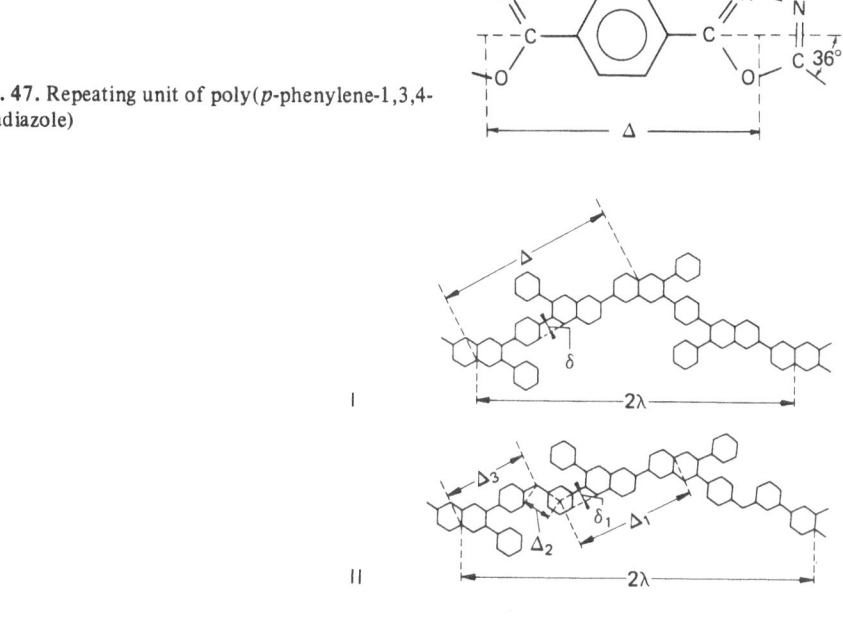

Fig. 47. Repeating unit of poly(p-phenylene-1,3,4-oxadiazole)

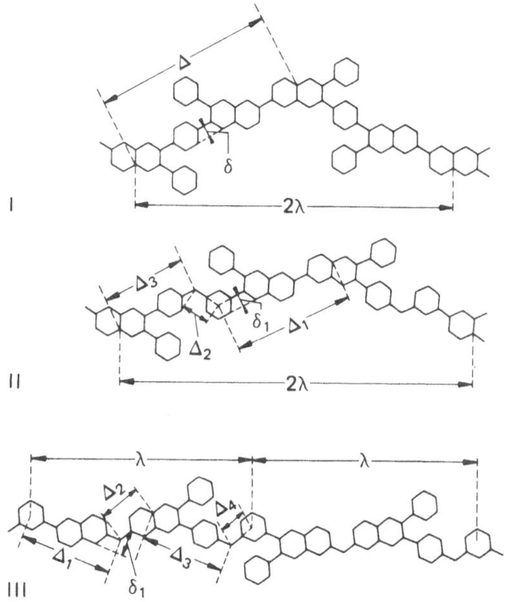

Fig. 48. Extended conformations of polyphenylquinoxaline molecules differing in the number of rotating bonds in the repeating unit I: one bond; II: three bonds; III: four bonds

The repeating unit of the poly(p-phenylene-1,3,4-oxadiazole) chain (Fig. 47)[47] contains one rotating bond $\Delta \approx 7$ A forming an angle $\vartheta \approx 36°$ with the bond of the neighbouring unit. In this case in order to calculate S_f and A_f in Eq. (68), δ should be taken equal to 0. A comparison of A_f with the experimental value, A, (Table 9) yields σ given in Table 11.

Similar calculations of S_f and A_f have been carried out for polyphenylquinoxalines[156] (Fig. 48). The values of σ obtained by a comparison with the experimental values of $\Delta n/\Delta \tau$ and A are also presented in Table 11.

For all ring-containing aromatic polymers the parameters of hindrance to rotation are close to each other and are lower than the values which are characteristic of typical flexible-chain polymers[10]. This means that, in contrast to cellulose derivatives, the high equilibrium chain rigidity of these aromatic polymers is caused by the peculiarities of the geometrical structure of their molecules rather than by hindrance to intra-molecular rotation.

It should be remembered that the relatively low value of the σ parameter obtained experimentally in the study of equilibrium properties does not necessarily mean that the rotation is similar to "true free rotation". In principle, it can correspond to the existence of high potential barriers U hindering rotation if the potential hindrance curve U (φ) is relatively symmetrical. Under these conditions, the molecule of a *para*-aromatic polyamide should behave as a kinetically rigid system modelled by a crankshaft with an array of rigid crankshaft parts uniformly distributed along azimuths of rotation φ. However, even in those cases where these rotations are "truly free", they cannot cause a marked deviation of the shape of *para*-aromatic polymer molecules from rod-like shape since the axes of these rotations are approximately parallel. Hence, the molecule of a *para*-aromatic polyamide retaining its elongated shape behaves as a kinetically rigid chain regardless of the degree of hindrance to intramolecular rotations.

4.5.4 Poly(Cyclohexane Amide)s

The elongated shape of the molecule can also be retained if aromatic rings are replaced by other groups, provided these groups ensure approximately paralled directions of rotational axes in the chain. Dynamo-optical properties of poly(cyclohexane amide) (PCHA)[157], a hydrated analog of PPBA or PPPhTPhA, confirm this assumption. In PCHA phenyl rings are replaced by cyclohexyl rings but rotating bonds in the chain remain parallel. Consequently, the molecule retains the "crankshaft" structure in which the chain size in the direction parallel to rotating axes increases proportionally to molecular weight regardless of whether the chain assumes an "extended" or a "coiled" conformation (Fig. 49). Accordingly, the shear optical coefficient $\Delta n/\Delta \tau$ is more than six times higher in solutions of PCHA sulfuric acid than in solutions of polycaprolactam, a linear (acyclic) analog of PCHA (Table 10) although the anisotropy of the monomer unit of the latter is twice as high. The use of these data and Eq. (71) yields A which, for PCHA molecules, is higher by an order of magnitude than for polycaprolactam (Table 10).

The copolymerization of PCHA and caprolactam can yield copolymers with a flexibility intermediate between those of the two components. In this case, just as for aromatic polyamides (Fig. 44), the rigidity (number of monomer units S per segment) of the copolymer can be evaluated a priori from the relative composition Z and rigidities S_1 and S_2 of the components by using the rule of the additivity of flexibilities (Eqs. (74) and (75)). This is demonstrated in Fig. 50 which shows that minute amounts of a more flexible component can greatly reduce chain rigidity.

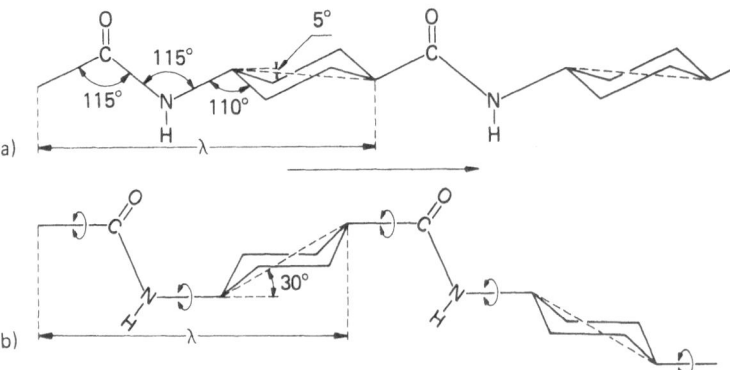

Fig. 49 a, b. Conformation of a poly(cyclohexane amide) chain: **a** extended, **b** coiled

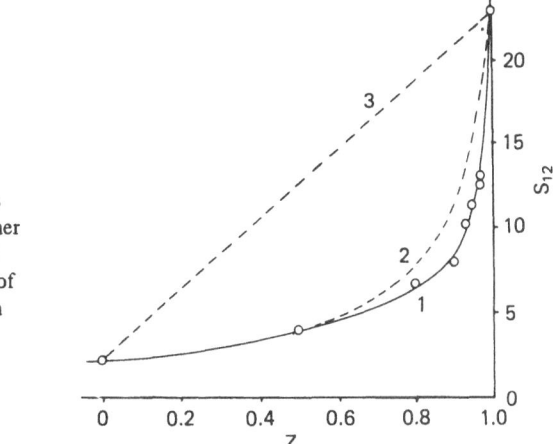

Fig. 50. Number of monomer units S_{12} in a Kuhn segment of copolymer molecules of cyclohexanamide and caprolactam vs. relative content Z of cyclohexanamide units in the chain according to flow birefringence data [157]. Points and Curve 1: experimental data; Curves 2 and 3: plotted according to Eqs. (74) and (75)

4.6 Comb-Like Polymers

Polymer molecules with long-chain side groups exhibit some characteristic properties many of which can be detected quantitatively by the method of flow birefringence.

Usually, these molecules are called "comb-like". Although side chains of the molecules contain the major part of their mass, the properties of the molecule as a whole in solution are determined not only by side chains but also by the main chain.

4.6.1 Poly(Alkyl Acrylate)s and Poly(Alkyl Methacrylate)s

The effect of the length and structure of side groups on the conformation of polymer molecules and the equilibrium rigidity of the main chain has been studied systematically for homologous series of poly(alkyl acrylate)s

$$\cdots -CH_2-CH- \cdots$$
$$|$$
$$CO-O-C_mH_{2m+1}$$

up to poly(octadecyl acrylate) (m = 18), poly(alkyl methacrylate)s

$$CH_3$$
$$|$$
$$\cdots -CH_2-C- \cdots$$
$$|$$
$$CO-O-C_mH_{2m+1}$$

up to poly(cetyl methacrylate) (m = 16) and also for some other comb-like polymers (Table 12).

Table 12. Values of the Kuhn segment A, segmental anisotropy $\alpha_1 - \alpha_2$, and anisotropy of the monomer unit Δa of some comb-like macromolecules

Polymer	Monomer unit	m^a	A (Å)	$(\alpha_1 - \alpha_2)$ $\times 10^{25}$ cm^3	$\Delta a \times 10^{25}$ cm^3
Poly(acrylic acid ester)s	$\cdots -CH_2-CH-\cdots$ $\quad\quad\quad\quad$ $C{=}O$ $\quad\quad\quad\quad$ O $\quad\quad\quad\quad$ C_mH_{2m+1}	1	21	+16	+1.9
		4	23	−17.8	−1.9
		8	39	−47.9	−3.6
		10	50	−74	−3.7
		12	40	—	—
		16	56	−141	−6.4
		18	72	−190	−6.6
Poly(methacrylic acid ester)s	$\cdots -CH_2-\overset{CH_3}{\underset{}{C}}-\cdots$ $\quad\quad\quad\quad$ $C{=}O$ $\quad\quad\quad\quad$ O $\quad\quad\quad\quad$ C_mH_{2m+1}	1	17	+2	+0.3
		4	17	−14	−2.1
		6	21	−40	−4.6
		8	20	−47	−5.9
		12	30	—	—
		16	44	−160	−8.9
Poly(cetyl vinyl ether)	$\cdots -CH_2-CH-\cdots$ $\quad\quad\quad\quad$ O $\quad\quad\quad\quad$ $C_{16}H_{33}$		60	—	—
Poly(cholesteryl acrylate)	(cholesteryl acrylate structure)		60	−360	−15
Poly[p-(p-cetylbenzoyloxy)phenyl methacrylate]	(cetylbenzoyloxy phenyl methacrylate structure with $C_{16}H_{33}$)		61	−3100	−130

a m = number of side chain carbon atoms

Sedimentation, diffusion, viscometric and flow birefringence measurements in dilute polymer solutions have been applied for this purpose[158−160].

The experimental data shown in Fig. 51 and Table 12 indicate that an increase in the length of side groups leads to an increase in the equilibrium rigidity of the main chain. This may be due to side group interactions. However, the length of the side chains of these polymers differs only by a factor of 2−4 at the utmost so that the main chain retains its flexibility virtually corresponding to that of typical flexible-chain polymers.

Hence, the ratio $\Delta n/\Delta \tau$ in solutions of these polymers is essentially independent of molecular weight[158]. Hence Eq. (67) may be used for the determination of $\alpha_1 - \alpha_2$.

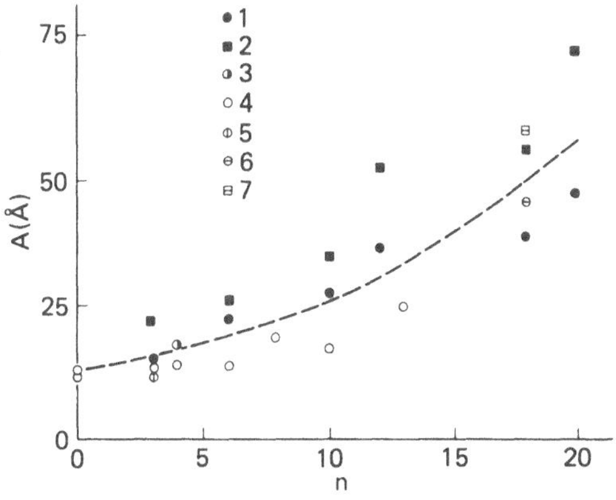

Fig. 51. Graph of Kuhn segment A of the main chain vs. number of valence bonds n in the side chain of poly(alkyl acrylate)s (1−3) and poly(alkyl methacrylate)s (4−7) according to viscometric (1, 3−6) and translational friction (2, 7) data

Data in Table 12 also show that an increase in the length of side chains causes a lowering of the positive segmental anisotropy of the molecule, $\alpha_1 - \alpha_2$, a change in sign and an increase in the negative value. Similar experimental data have been obtained for poly(1-alkenes)[167]. The change in sign indicates that for higher homologs of these polymers the main contribution to the anisotropy of the monomer unit Δa is provided by the side group whose polarizability in the direction normal to the main chain is much higher than that in the direction of the main chain.

The dependence of the value and sign of Δa on the length of the side group can be utilized to study the anisotropy and flexibility of side chains of comb-like molecules. For this purpose, the side group should be described by a worm-like chain the direction of which at the beginning (axis Z in Fig. 14) is normal to that of the main chain.

Then, according to Eq. (39), the contribution of the side group to the anisotropy of the monomer unit of the comb-like chain is given by

$$\Delta b_L = -S^* \cdot \Delta a^* \cdot (1 - e^{-6 n/S^*})/12 \tag{77}$$

where n and S^* are numbers of valence bonds in the side chain and in its segment, respectively, and Δa^* is the anisotropy of a part of the side chain containing one valence bond.

The theoretical dependence described by Eq. (77) is compared with experimental data obtained for poly(alkyl methacrylate)s[158] and poly(alkyl acrylate)s[159] (Figs. 52 and 53). Points represent experimental values given in Table 12. Solid curves depict theoretical dependences according to Eq. (77) plotted for values of parameters Δa^* and S^* that ensure the best agreement

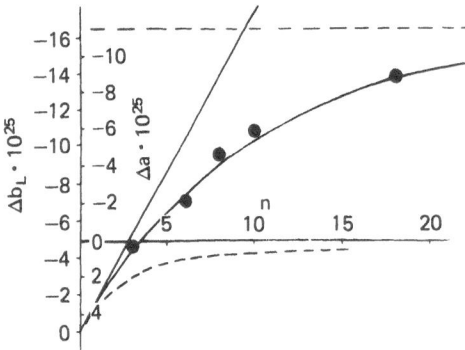

Fig. 52. Anisotropy of monomer unit Δa and contribution of the side chain to this anisotropy, Δb_L, for poly(alkyl methacrylate)s in benzene vs. number of valence bonds n in the side chain. Solid curve: theoretical dependence according to Eq. (77) at $\Delta a^* = 3.3 \cdot 10^{-25}$ cm^3 and S$^* = 60$; broken curve: plotted according to Eq. (77) at $\Delta a^* = 3.3 \cdot 10^{-25}$ cm^3 and S$^* = 16$, corresponding to the properties of free polymethylene chains in solutions

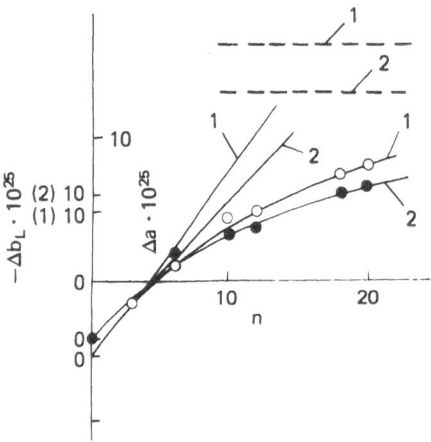

Fig. 53. Anisotropy of monomer unit Δa (1 = toluene, 2 = decalin) vs. the number of valence bonds n in the side chain. Solid curves: theoretical dependence according to Eq. (77): $1 : \Delta a^* = 2.6 \cdot 10^{-25}$ cm^3, S$^* = 80$; $2 : \Delta a^* = 2.1 \cdot 10^{-25}$ cm^3, S$^* = 80$

between the theoretical curve and experimental points. The values of Δa^* range from 2.1 to $3.3 \cdot 10^{-25}$ cm^3 which is in agreement with the experimental value known for the polymethylene chain[67]. In contrast, the value of S* is 60 for poly(alkyl methacrylate)s and 80 for poly(alkyl acrylate)s which is four to six times higher than the values of S for free polymethylene (polyethylene) chains in solution[6]. High equilibrium rigidity of side chains of these polymers can be ascribed to interactions between these chains, their distances inside the comb-like molecule being much smaller than those between free chain molecules in dilute solution.

The analysis of experimental data on flow birefringence in solutions of poly-(1-alkene)s with different lengths of the side chain leads to a similar conclusion[11].

4.6.2 Molecules Containing Mesogenic Side Groups

Interactions between side chains leading to unique dynamo-optical properties are particularly pronounced for comb-like molecules with mesogenic side groups.

Poly(phenyl methacrylic esters) of p-alkoxybenzoic acids (PPhEAA) are an example of this type of polymers

$m = 3, 6, 9, 16$

The main part of the side group in these macromolecules consists of the alkoxybenzoic acid moiety. This acid may form thermotropic liquid crystals. The investigation of the hydrodynamic properties of PPhEAA molecules in dilute solutions[163] has revealed that the equilibrium rigidity of their main chains is relatively low (Table 12). Hence, since for all flexible-chain polymers, the shear optical coefficient $\Delta n/\Delta \tau$ in PPhEAA solutions is independent of molecular weight the segmental anisotropy $\alpha_1 - \alpha_2$ and the anisotropy of the monomer unit Δa may be determined by use of Eq. (67).

Negative segmental anisotropy $\alpha_1 - \alpha_2$ is higher by more than an order of magnitude than that of poly(cetyl methacrylate) (Table 12). It is only comparable to the corresponding values of such rigid polymers as poly(alkyl isocyanate) or *para*-aromatic polyamides (Table 9). The anisotropy of PPhEAA is higher than that of poly(p-benzamide) and higher by more than an order of magnitude than that for poly(butyl isocyanate). This implies that high values of $\Delta n/\Delta \tau$ and $\alpha_1 - \alpha_2$ for PPhEAA molecules are due to interactions of their side groups rather than to the rigidity of their main chain. As a result, interactions of PPhEAA side chains become orientationally ordered, the degree of ordering being close to that of the nematic mesophase.

4.6.3 Graft Copolymers

Macromolecules of styrene-methyl methacrylate graft copolymers are another example of comb-like molecules; they have been extensively studied by flow birefringence[84, 115, 164, 165].

The optical properties of these polymers are very peculiar. Thus, the optical anisotropy of graft copolymers can exceed by many times that of the grafted homopolymers. Moreover, it is positive, i.e. it is opposite in sign to the negative anisotropy of the polystyrene chain although the polystyrene mass in the graft copolymer exceeds 90% whereas the absolute value of the negative segmental anisotropy of the styrene homopolymer chain is higher by over an order of magnitude than the positive segmental anisotropy of poly(methyl methacrylate)[67].

This unusual combination of the optical properties of the grafted components with those of the graft copolymer can be explained by taking into account the specific molecular structure of the latter (Fig. 54).

The negative anisotropy of a single polystyrene molecule is determined by the predominant orientation of the planes of phenyl rings normal to the main chain. The positive anisotropy of the graft copolymer molecule implies that the correlation between the orientation of phenyl rings in the side groups is spread to the whole copolymer molecule. In each monomer unit of the copolymer the planes of the phenyl rings are predominantly oriented parallel to the main chain (Fig. 54, axis 1) whereas inside the entire copolymer molecule they are parallel to the axis joining the ends of the main chain.

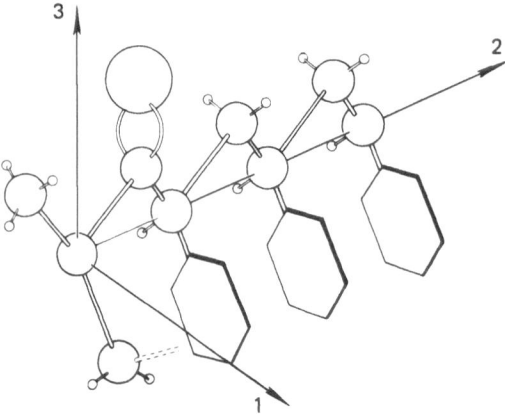

Fig. 54. Monomer unit of a molecule of methyl methacrylate-styrene graft copolymer. Axis 1: parallel to the *trans* chain of poly(methyl methacrylate) (i.e. the main chain of the graft copolymer the plane of which is the 1,3 plane). Axis 2: parallel to the direction of the *trans* chain of grafted polystyrene

All these properties are observed under the condition that the side chains of the graft copolymer are much shorter than the main chain, i.e. the molecules must exhibit a comb-like structure. Moreover, the higher the degree of grafting, i.e. the smaller the distance between the chains of grafted polystyrene, the more pronounced these properties. This shows that the specific properties of graft copolymer molecules, just as those of the comb-like molecules previously considered (see p. 161), are due, to a great extent, to the interactions of their side chains.

The application of the optical anisotropy theory of comb-like molecules to graft copolymers permits the quantitative determination of the optical anisotropy and the flexibility of both the main and side chains by using Eq. (77)[84, 165].

These data show that the length of the Kuhn segment A in both the main and side chains of the graft copolymer can be higher by over an order of magnitude than that of the components being copolymerized. As a result, a marked dependence of shear optical coefficient $\Delta n/\Delta \tau$ on the length of the main chain[114, 165] which is characteristic of rigid-chain polymers, can be observed in graft copolymer solutions.

5 Electric Birefringence

Electric birefringence (EB) or the Kerr effect is widely used in molecular optics as a method for the investigation of the molecular structure of low molecular weight substances. The study of the Kerr effect in the gas phase or in solutions in combination with other methods, such as refraction, light scattering, dielectric measurements etc. permits to ascertain the spatial arrangement of atoms in the molecule and thus to calculate the main values of the polarizability tensor of the molecule and to obtain information about the value and direction of its dipole moment[4, 166].

The Kerr constant may serve as the characteristic value of birefringence introduced by the dissolved substance

$$K \equiv (\Delta n/cE^2)_{\substack{E \longrightarrow 0 \\ c \longrightarrow 0}}$$ (78)

where $\Delta n = n_p - n_s$ is the excess difference (obtained by subtracting the effect in the pure solvent from that in solution) between refractive indices of two light beams with the directions of vibrations of the electric vector parallel (n_p) and normal (n_s) to the electric field. E is the intensity of the electric field the direction of which is normal to the light beams and c is the solvent concentration.

The possibility of using the Kerr effect for the study of the structure and conformation of polymer molecules greatly depends on whether it is used for solutions of flexible-chain or rigid-chain molecules[167].

5.1 Flexible-Chain Polymers

The study of the Kerr effect in polystyrene solutions has shown[168, 169] that the observed EB does not differ either in sign or in value from that of the solution of the corresponding monomer of equal concentration and is independent of the molecular weight of the polymer. Similar results have been obtained in the investigation of EB in solutions of many linear carbon-chain polymers[170–172]. Sometimes, the values of K vary for polymers differing in molecular weight[171, 172]. This is attributed either to different chain tacticity[173] or to the excluded volume effect[174]. However, both points of view are doubtful since usually the differences in K observed in a series of molecular weights are small, approach experimental errors and are poorly reproducible.

The pronounced dependence of the Kerr constant on molecular weight observed for poly(vinyl chloride)[175] and poly(vinyl bromide)[176] solutions appears to be due to their molecular inhomogeneity. The presence of aggregates in solutions of these two polymers has often been detected experimentally[177–179].

The theory of the Kerr effect in the steady-state electric field was put forward by Peterlin and Stuart[4, 180] for flexible polymer molecules described by a chain of freely jointed Kuhn segments. This theory is based on the concept of mutually independent orientations of segments in the electric field. Consequently, it result coincides with the Langevin-Born theory for low-molecular weight gases and liquids[181, 182] if the values of the dipole moment and the optical and dielectrical polarizabilities in these equations are assumed to be related to one segment of the polymer molecule. Hence, birefringence in the Peterlin-Stuart theory is proportional to the overall concentration of segments in solution regardless of whether they are distributed in long- or short-chain molecules. Therefore, the main aspect of the theory is the conclusion that the Kerr constant K is independent of the molecular weight of the polymer. This conclusion also remains valid for the theories that do not use the model of a freely jointed chain but relate K directly to such structural parameters of the molecule as polarizabilities and dipole moments of chemical bonds and to the type of rotation of these bonds in the polymer chain[8, 173, 174, 183]

The above-mentioned experimental data agree qualitatively with the conclusion of the theory that K is independent of molecular weight. However, the fact that the value of K for the solution of a flexible-chain polymer does not exceed that of K for the solution of the corresponding monomer and is sometimes lower than the latter[171, 172, 175, 176] implies that the anisotropy (and the size) of a kinetic unit oriented more or less independently in the electric field is lower than that of the Kuhn segment; the Kuhn segment contains 6−8 monomer units for most flexible-chain polymers[10]

Only a small number of researchers have attempted to interpret the experimental data on the Kerr effect from the standpoint of theories based on a rotational isomer mechanism of chain flexibility and tensor additivity of the polarizability of chain bonds[8, 173, 174, 184]. These attempts did not yield good agreement between theory and experimental data. This may result from both the unreliability of the data and the inadequacy of some suggestions used in the theory (additivity of optical properties of valence bonds, character of the internal field etc.).

The experimental data on the non-steady state Kerr effect of flexible-chain polymers dissolved in solvents with moderate viscosities reveal that at frequences up to 10^7 Hz no dispersion of EB is observed (just as in solutions of low molecular weight substances and monomers). This is also an indication of mutually independent orientation of single monomer units in the electric field which is only slightly related to the structure and conformation of the polymer chain as a whole.

Hence, as far as solutions of flexible-chain polymers are concerned, at present the Kerr effect cannot be considered as an effective method for the study of conformational and structural characteristics of polymer molecules.

The main difficulty in the experimental investigations of EB in flexible-chain polymer solutions is due to the low value of the observed effect. Specific Kerr constants for a flexible-chain polymer bearing no charge are $K \approx 10^{-12}$ cm^5 g^{-1} $(V/300)^{-2}$ even for polar macromolecules and, hence, the difference between birefringence in a dilute solution and the Kerr effect in the solvent alone is very slight.

5.2 Polyelectrolytes

The situation is more favourable for the study of EB in solutions of flexible-chain polyelectrolytes for which the value of K may be higher by several orders of magnitude than for molecules bearing no charge[185−188]. This seems plausible since the uncoiling of a flexible-chain polyion by electrostatic repulsion of ionogenic groups increases the persistent length of the chain and the optical and hydrodynamic properties of the molecule approach those of a rigid-chain polymer[67, 189−191] (Table 8). This permits the interpretation of experimental data by using the electrooptical properties of flexible-chain polymers in terms of a worm-like chain model[192−194]. However, EB in solutions of polyelectrolytes is of a complex nature. The high value of the observed effect is caused by the polarization of the ionic atmosphere surrounding the ionized macromolecule rather than by the dipolar and dielectric structure of the polymer chain. This polarization induced by the electric field depends on the ionic state of the solution and the ionogenic properties of the polymer chain whereas its dependence on the chain structure and conformation is slight. Hence, the information on the optical, dipolar and conformational properties of macromolecules obtained by using EB data in solutions of flexible-chain polyelectrolytes is usually only qualitative. Studies of the kinetics of the Kerr effect in polyelectrolytes (carried out by pulsed technique) are more useful since in these systems relaxation phenomena are observed; they can be utilized to obtain information on the rotational mobility of the polymer chain[185, 192–194].

Numerous data are available on the Kerr effect in solutions of ionogenic chain molecules with secondary structures[195–197], particularly of polypeptides displaying a helical conformation[198, 199]. A high EB value in these solutions is ensured by both the polarization of ionic atmospheres and the character of the rigid secondary structure of the molecules. It is not always possible to distinguish between the effect of the structural order and the ionic state of the medium on the electro-optical properties of rigid polyelectrolyte solutions. Thus it is difficult to determine their molecular structure.

5.3 Chain Molecules with Rigid Secondary Structures

It is possible to obtain more reliable quantitative information on the conformational and structural characteristics of polymer chains by studying EB in solutions of polypeptides that do not contain ionogenic groups and using non-electrolytic spiralizing systems as solvents[199–207].

Poly(γ-benzyl-L-glutamate) (PBLG) is a typical representative of these polymers. Its solutions in spiralizing solvents exhibit very high positive birefringence both in mechanical (Table 9, Fig. 33) and electric fields.

The very high equilibrium rigidity of PBLG in the helical conformation (Table 9), resulting in high values of $\Delta n / \Delta \tau$ and of the Kerr constant K, drastically decreases in solutions during the conformation helix-coil transition. This can be seen in Fig. 55 which shows the dependence of K on the percentage of a de-spiralizing component (dichloroacetic acid) in mixed solvents[204]. The addition of this acid to the solution leads to the breaking of intramolecular hydrogen bonds and induces the conformational transition. As a result, PBLG loses its helical conformation and becomes a typical flexible-chain polymer with the length of the Kuhn segment of 20 Å. The conformational transition shown in Fig. 55 clearly indicates that a molecule with the same chemical structure exhibits widely different values of electro-

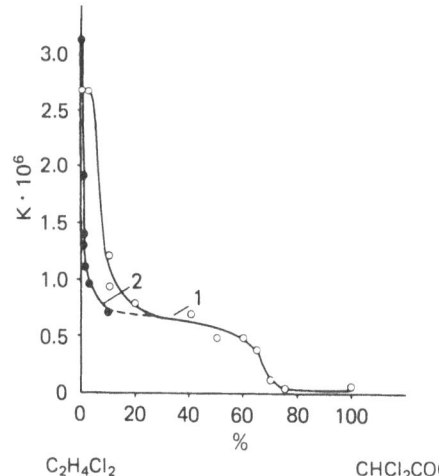

Fig. 55. Kerr constant K vs. dichloroacetic acid (DChAA) content in solutions of the following systems: 1: PBLG – dichloroethane (DChE) – DChAA – dimethylformamide; 2: PBLG – DChE – DChAA[204]. K is expressed in $cm^5 g^{-1} (v/300)^{-2}$ units

optical effects in helical and in coil conformations: during the helix-coil transition
K for a polymer with M = 3.3 x 10^5 changes from 2.7 x 10^{-6} to 3 x 10^{-9} cm^5 g^{-1}
$(V/300)^{-2}$.

Another important difference in the electro-optical properties of rigid-chain
and flexible-chain polymers is shown by the dependence of the Kerr constant K on
molecular weight M:

For flexible-chain molecules K is virtually independent of M. In contrast, for rigid-chain
polymers K usually increases with M (Fig. 56)[202]. Although in both solvents the conformation of PBLG molecules is helical and equally rigid, the value of K in dichloroethane solutions
is twice as large as that in m-cresol. This is caused by a high refractive index increment dn/dc in
the former solvent and the correspondingly greater contribution of the microform effect[202,204]
to the optical anisotropy of the PBLG molecule (Eq. (50)) in dichloroethane.

Finally, the third significant property of rigid-chain polypeptides distinguishing them from
flexible-chain polymers is the dispersion of the Kerr effect observed in their solutions (in spiralizing solvents) in the range of radio frequencies of the alternating field (Fig. 57)[201]. With increasing ν, EB decreases almost to zero. The range of dispersion is sharply shifted toward high ν
with decreasing molecular weight of the polymer. This frequency dependence of EB reveals that
relaxation phenomena exist in the process of orientation of PBLG molecules and that this orientation is of a dipolar character.

All these features of EB in solutions of rigid helical molecules (high value of K,
dependence of K on M and frequency dependence of EB) are more or less typical

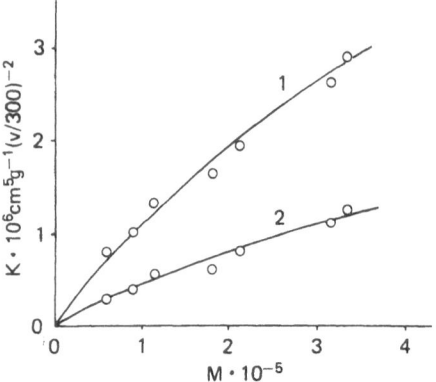

Fig. 56. K vs. M plot for solutions of PBLG
in dichloroethane (1) and m-cresol (2)[203]

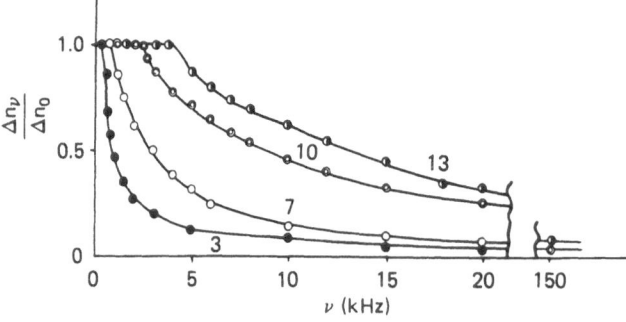

Fig. 57. Relative EB value $\Delta n_\nu/\Delta n_0$ for solutions of PBLG fractions in dichloroethane vs. frequency ν of the applied sinusoidal field[202]. Curves 3, 7, 10 and 13 refer to polymers with different molecular weights: M = 3.2 · 10^5 (3); 1.4 · 10^5 (7); 0.9 · 10^5 (10); 0.6 · 10^5 (13). The
concentration ranges from 0.1 · 10^{-2} to 0.5 · 10^{-2} g/cm^3

of other rigid-chain polymers without secondary structure: polyisocyanates, poly-amides, cellulose esters, ladder and comb-like polymers etc. However, in contrast to polypeptides, the electro-optical properties of these polymers are determined by the peculiarities of the chemical structure of their molecules rather than by the specificity of the secondary structure. Hence, the study of the Kerr effect in non-ionogenic solutions of these polymers permits the establishment of a direct correlation between the EB of these solutions and the conformational and structural characteristics of the macromolecules.

5.4 Kinetics of the Kerr Effect in Solutions of Rigid-Chain Polymers

The study of the kinetics of EB in rigid-chain polymer solutions is of great importance since it permits the understanding of the physical nature of this phenomenon and provides information on the mechanism of molecular motion in the electric field.

If the polymer chain contains polar groups (dipole μ_0), then in the electric field each of these groups should exhibit a rotating moment orienting the dipole μ_0 in the field direction (Fig. 58a). If the rotations of single polar groups are relatively weakly correlated, the groups are oriented in the field more or less independently of each other. As a result of these intramolecular rotations (small-scale motion) in the electrical field the conformation of the molecule changes. i.e. it undergoes deformation.

However, in any conformation the chain as a whole has a dipole moment μ being the vector sum of dipole moments μ_0 of all its polar groups. The rotating moment exhibited by the dipole μ in the electric field, can lead to the orientation of μ in the field direction through rotation of the chain molecule as a whole (large-scale motion). In this case, the mechanism of the appearance of the polarization and macroscopic solution anisotropy may be called orientational mechanism.

As already mentioned (pp. 111 and 112) the deformational mechanism is characteristic of kinetically flexible molecules for which the orientation time τ_0 is greater than the deformation time τ_d, i.e. the time during which the molecule in solution retains a random conformation (relaxation time of the conformation).

In contrast, the motion in the electric field of kinetically rigid molecules for which τ_0 is less than τ_d should be governed by the mechanism of the orientation of the molecule as a whole and the rate of this process should be determined by the rotational friction coefficient W or rotatory diffusion coefficient D_r (Eq. (53)). Hence, the study of the orientation kinetics of kinetically rigid macromolecules in the electric field permits the determination of their rotational mobility (i.e. W and D_r).

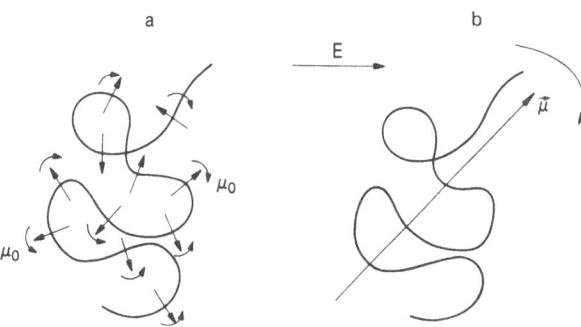

a   b

E

Fig. 58a, b. Possible polarization mechanisms of a chain molecule in the electric field. a Kinetically flexible chain (deformational mechanism); b kinetically rigid chain (orientational mechanism)

5.4.1 EB Dispersion

For this purpose, the Kerr effect in the alternating (sinusoidal) field can be used. The theory of this effect for solutions of kinetically rigid molecules has been formulated by Peterlin and Stuart[209]. According to this theory, the character of the dependence of the observed birefringence Δn_ν on frequency $\nu = \omega/2\pi$ of the applied field clearly differs for the two cases:

1) Orientation of the macromolecules results from the anisotropy of their polarizability

$$\Delta n_\nu = \Delta n_0 \left[1 + \frac{\cos(2\omega t - \delta_1)}{\sqrt{1 + \omega^2 \tau_1^2}} \right] \tag{79}$$

2) Orientation is due to the dipole moment of the macromolecules

$$\Delta n_\nu = \Delta n_0 \left[\frac{1}{1 + \omega^2 \tau_2^2} + \frac{\cos(2\omega t - \delta_2)}{\sqrt{1 + \omega^2 \tau_2^2} \cdot \sqrt{1 + \omega^2 \tau_1^2}} \right] \tag{80}$$

where $\tau_1 = 1/3 \, D_r$ and $\tau_2 = 1/2 \, D_r$ are relaxation times of the effect caused by the anisotropy of the polarizability and the permanent dipole of the molecule, respectively.

According to Eq. (79) when a technique recording the values of Δn_ν, averaged over the period of the alternating field, is applied (e.g. a visual technique), Δn_ν is independent of frequency and equal to Δn_0, i.e. to EB in the steady-state field. In contrast, according to Eq. (80), the dipolar part of the Kerr effect is characterized by the dispersion which can be described by

$$\Delta n_\nu = \Delta n_0/(1 + \omega^2 \tau^2) \tag{81}$$

with good approximation. Eq. (81) is analogous to the well-known Debye's equation describing the dispersion of the dielectric polarization in a solution of dipolar molecules.

Relaxation time τ in Eq. (81) is related to the rotatory diffusion coefficient D_r with respect to the axis normal to the dipole of the molecule

$$\tau = 1/2 \, D_r \tag{82}$$

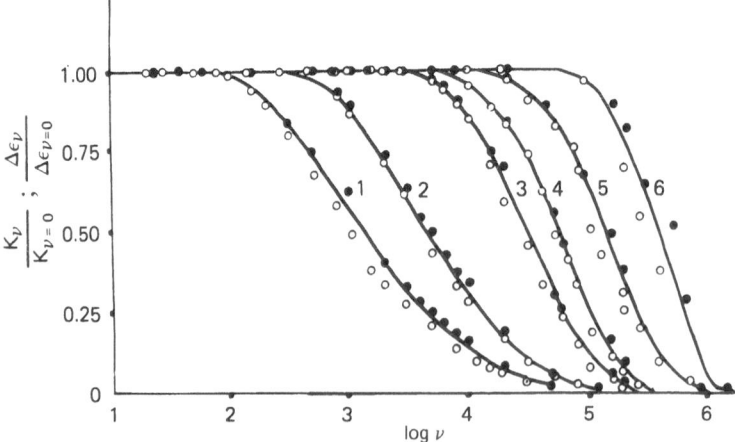

Fig. 59. Relative Kerr constant $K_\nu/K_{\nu = 0}$ (open circles) and relative dielectric increment $\Delta \epsilon_\nu/\Delta \epsilon_{\nu=0}$ (filled circles) vs. frequency of the electric field ν for poly(chlorohexyl isocyanate) fractions in dioxane: $M \cdot 10^{-4} = 27.4$ (1); 12.5 (2); 3.7 (4); 2.3 (5); 1.0 (6)[12, 210−212])

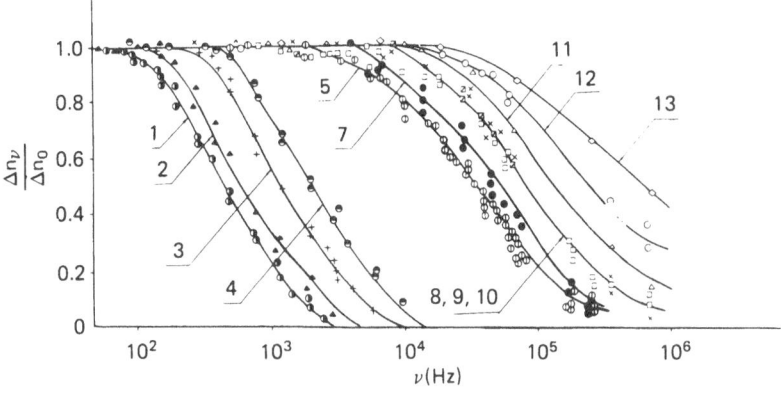

Fig. 60. $\Delta n_\nu / \Delta n_0$ vs. ν plot for cellulose carbanilate fractions in dioxane. Curves correspond to the following molecular weights: $M \cdot 10^{-3}$ = 870 (1); 680 (2); 440 (3); 280 (4); 84 (5); 66 (7); 44 (8); 46 (9); 39 (10); 34 (11); 28 (12); 24 (13)[213-215]

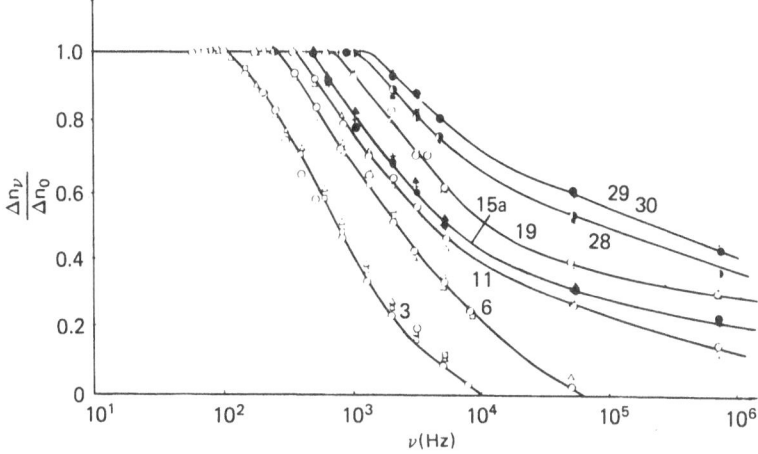

Fig. 61. $\Delta n_\nu / \Delta n_0$ vs. ν plot for solutions of ladder polychlorophenylsiloxane in benzene. Number of curves refer to the following molecular weights: $M \cdot 10^{-3}$ = 1200 (3); 720 (6); 550 (11); 400 (15a); 310 (19); 120 (28); 110 (29); 88 (30)[216]

As already indicated, (p. 170) the dispersion of the Kerr effect in the range of radio frequencies is a characteristic property of rigid-chain polymer solutions. This can be seen in Figs. 59–61 which show frequency dependences of EB for solutions of poly(chlorohexyl isocyanate)s, cellulose carbanilate and ladder polychlorophenyl-siloxane. Similar dependences have been obtained for poly(butyl isocyanate)[217], various cellulose ethers and esters[218] and ladder polysiloxanes[219, 32].

As frequency increases, all dispersion curves decrease to virtually zero. This indicates that EB is produced by the dipolar-orientational mechanism whereas the anisotropy of the dielectric polarizability of the macromolecules is virtually imper-ceptile in the Kerr effect. The displacement of curves towards higher frequencies with decreasing M means that relaxation times τ increase with rising M. These properties are a direct proof that the rotation of the macromolecule as a whole is the mechanism of molecular motion responsible for EB. This may be interpreted as the manifestation of the kinetic rigidity of macromolecules.

In the same frequency range, the dispersion of the dielectric increment $\Delta\epsilon$ of solution can be observed. The curves of dielectric dispersion almost coincide with the dispersion curves for EB (Fig. 59) so that both mechanisms of molecular motion are identical and are represented by dispersion Eq. (81) for kinetically rigid molecules.

5.4.2 Relaxation Time, Molecular Weight and Viscosity

Relaxation times τ obtained from experimental dispersion curves in Figs. 59–61 by using Eq. (81) can be compared with molecular weights M of the same polymer samples. The dependence of τ on M for poly(butyl isocyanate) is shown in Fig. 62. The experimental points fit a curve the slope of which decreases with increasing M from 2.7 to 1.5. Taking into account Eq. (82) this can be expressed by

$$D_r = bM^{-n} \tag{83}$$

where n decreases from 2.7 to 1.5 with rising M. This relationship corresponds to the concept that the kinetic unit rotating in the electric field is the polar molecule as a whole and that with increasing. M its hydrodynamic properties change from those of a straight rod to those of the undrained Gaussian coil.

Comparison of relaxation times with intrinsic viscosities $[\eta]$ of the same samples allows to draw similar conclusions. As we could see, the theoretical relationship between the rotational friction coefficient for kinetically rigid particles $W = kT/D_r = 2\,kT\,\tau$ and the product $M\,[\eta]\eta_0$ is given by Eq. (18) (p. 112). Here, coefficient F changes from 0.13 to 0.42 when the shape of the particle varies from the rod-like to the spherical shape. For the Gaussian coil, F has an intermediate value. Figure 63 shows the values of F calculated according to Eq. (18) by using the experimental values of τ, M, $[\eta]$ and η_0 for fractions of poly(chlorohexyl isocyanate)[210]. The

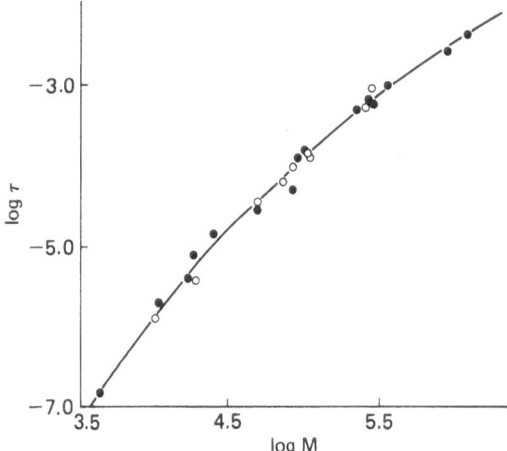

Fig. 62. τ vs. M graph in the Kerr effect (open circles) and dielectric polarization (filled circles) for solutions of poly(butyl isocyanate) fractions in tetrachloromethane[12]

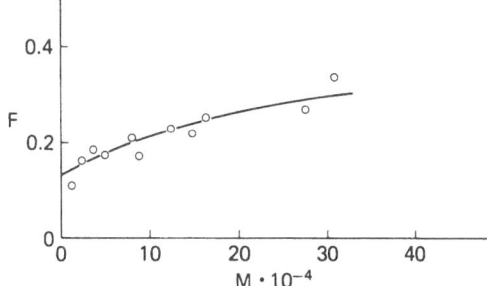

Fig. 63. F vs. M plot for poly(chlorohexyl isocyanate) fractions in tetrachloromethane[210]

values of F are not only within the range predicted by the theory of kinetically rigid particles but also clearly demonstrate the decrease in F with M. This is in accordance with the simultaneous change in the shape of the molecule from the random coil to the rod. Hence, the character of the molecular motion responsible for EB is similar to that of the motion accounting for the viscosity of the solution. It is the rotation of the molecule as a whole in the electric field or in the field of mechanical shear forces. Comparison of τ and $[\eta]$ for other rigid-chain polymers leads to a similar result[213, 216].

5.5 Rotational Friction and Geometrical Parameters of Chain Molecules

The orientational mechanism of EB in solutions of rigid-chain polymers and the possibility of determining rotatory diffusion constants of their molecules from dispersion curves may be utilized for the characterization of equilibrium conformational properties of their chains. The theory of rotational friction of kinetically rigid molecules developed by Hearst making use of the statistics of worm-like chains[19] can be employed for this purposes. The results of this theory for the two limiting cases of molecular conformation referring to the slightly bent rod and the worm-like coil are expressed by Eqs. (27) and (28) (Sect. 2.3).

As an example of the use of Eq. (28) (p. 114), the data for fractions of cellulose carbanilate, the molecules of which can be represented by a partially drained worm-like coil, are plotted in Fig. 64. The dependence of the expression of the left hand side of Eq. (28) on $M^{0.5}$ is approximated by a straight line the slope of which yields the length of the Kuhn segment A and the intercept the hydrodynamic diameter of the chain d. The curves in Fig. 64 provide the values of A = 160 A and d = 6 A for the cellulose carbanilate chain.

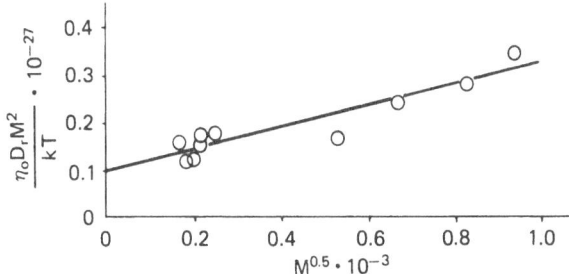

Fig. 64. $\eta_0 D_r M^2 / kT$ vs. $M^{0.5}$ graph for cellulose carbanylate fractions in dioxane[213, 214]

For the other limiting case, that of a slightly bent rod, the Hearst Equation (27)(p. 114) can be represented as[210]

$$\left[\frac{3\,kT}{\pi\eta_0 D_r}\left(\ln\frac{2\lambda M}{M_0 d}-\gamma\right)\right]^{\frac{1}{3}}\frac{M_0}{M}=\lambda\left\{1-\frac{M}{4\,M_0 S}\left[1-\left(\ln\frac{2\lambda M}{M_0 d}-\gamma\right)^{-1}\right]\right\} \tag{27a}$$

where

$$\gamma=1.57-7\left[\left(\ln\frac{2\lambda M}{M_0 d}\right)^{-1}-0.28\right]^2,$$

M_0 is the molecular weight of the monomer units, S the number of monomer units in a Kuhn segment and λ the length of the monomer unit in the chain direction.

Plotting the product on the left-hand side of Eq. (27a) versus M, λ is obtained from the initial ordinate of the curve and S from the initial slope.

This plot (Fig. 65) yields the values of S = 270 and λ = 2.3 A for poly(chlorohexyl isocyanate)[210] and S = 550 and λ = 2.10 A for poly(butyl isocyanate)[217].

The values of A, S, d and λ obtained from experimental data on the kinetics of the Kerr effect agree qualitatively with those determined for the same polymers by translational diffusion and sedimentation (Table 3). The agreement between geometrical molecular characteristics obtained from the phenomena of rotational and translational friction indicates that the hydrodynamic and the conformational models on which this theory is based are valid. This is an evidence of the kinetic rigidity of the investigated chains.

This conclusion applies equally to both helical polypeptides and poly(alkyl isocyanate)s on one hand and to ladder polysiloxanes and cellulose derivatives on the other hand. However, as discussed in Sect. 4.1 in the mechanical shear field the kinetic rigidity of polymers of the former group markedly exceeds that of the latter group. This difference in the evaluation of the kinetic flexibility of polymer molecules according to their dynamo-optical and electrical properties can be understood if the specificity of the electric structure of a polar chain molecule pointed out by Stockmayer[220–222] is taken into account. If monomer units of a chain molecule exhibit the component of a dipole uniformly directed along the main chain and rigidly attached to it, the sum of these components gives the total dipole moment of the molecule and the direction of the dipole moment coincides with the vector of the end-to-end distance h. It is formally equivalent to the presence of opposite charges at the chain ends. Hence, here the reorientation of mono-

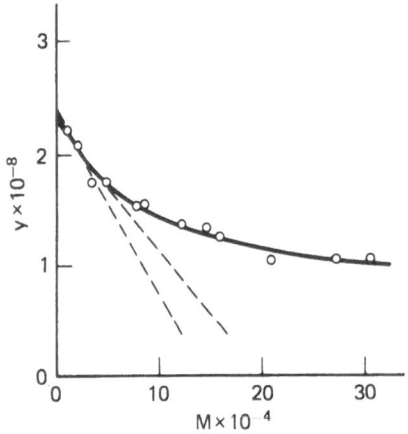

Fig. 65. Graph of parameter y = $\left[\frac{3\,kT}{\pi\eta_0 Dr}\left(\ln\frac{2\lambda M}{M_0 d}-\gamma\right)\right]^{1/3}\frac{M_0}{M}$ vs. M for poly(chlorohexyl isocyanate) fractions in tetrachloromethane[210]

mer dipoles in the electric field is inevitably related to the change in the mutual spatial arrangement of both chain ends, i.e. with large-scale motions in the molecule. In this molecule the cooperation of internal rotations in the electric field can occur owing to purely geometrical conditions of the dipolar structure of the chain rather than to the rigidity of its structure. This purely "polar" effect favouring the rotation of the molecule as a whole in the electric field is absent in the mechanical shear field of a laminar flow. Consequently, the same chain molecule, according to its kinetic properties, may seem to be less rigid in the mechanical shear field than in the electric field.

This discussion demonstrates that, in contrast to the equilibrium (static) rigidity, the concept of the kinetic rigidity of a chain molecule is not universal. The kinetic flexibility of the chain depends on the character of the process in which this flexibility is manifested.

5.6 Intramolecular Motions in the Electric Field

Although the orientation of macromolecules is the principal mechanism of molecular motion in rigid-chain polymer solutions, solution polarization by the molecular-deformational mechanism can frequently be observed. Figure 66 shows as an example the dependence of a dielectric increment on the frequency for solutions of cellulose carbanilate fractions in dioxane. Dispersion curves fit the range of frequencies in which the orientational mechanism of polarization of molecules with a given molecular weight cut off. When ν increases further these curves form a plateau the height of which is independent of molecular weight. This means that for all fractions the residual polarization of the solution in the plateau region is caused by a relaxation mechanism which can occur only at higher frequencies than those used in this experiment. Since in the plateau region $\Delta \epsilon/c$ is independent of M, this mechanism is evidently related to the small-scale intramolecular motion, i.e. to the deformational mechanism which is probably determined by the orientation of polar bonds of the molecule in labile side groups of the chain.

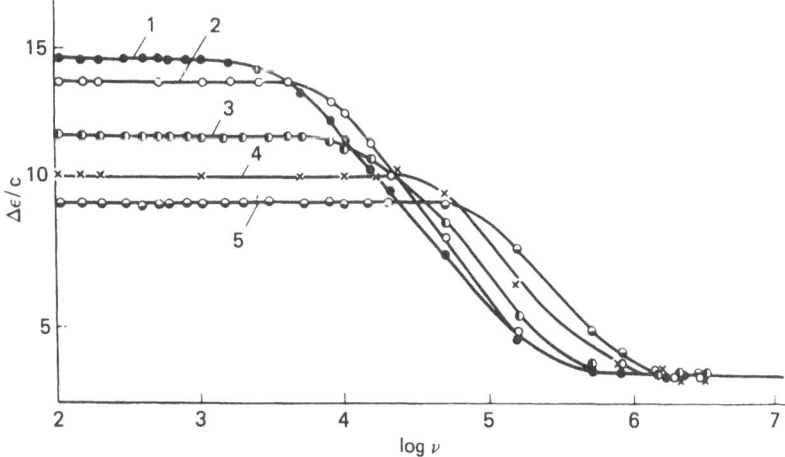

Fig. 66. Dielectric increment $\Delta \epsilon/c = (\epsilon-\epsilon_0)/c$ vs. ν plot for solutions of cellulose carbanilate in dioxane; ϵ_0, ϵ = dielectric permittivities of solvent and solution, respectively (at concentration c). $M \cdot 10^{-3} = 68$ (1); 58 (2); 43 (3); 28 (4); 23 (5)[223)]

However, this mechanism of motion does not provide any great contribution to the Kerr effect since the dispersion curves of EB fall to virtually zero (Fig. 60). This difference may be interpreted by the proportionality of the orientational EB of a rigid-chain polymer to the square of the number of monomer units in segment S^2 whereas increment $\Delta\epsilon/c$ related to the orientational mechanism is proportional to S (see Sects. 5.8 and 5.9). Hence, in the case of dielectric polarization the part played by the deformational mechanism as compared to the orientational mechanism can be more important than in the case of EB.

The display of intramolecular motion in rigid-chain polymers can also be observed if the kinetics of the Kerr effect is studied for a series of fractions of relatively high molecular weight. In fact, as already mentioned (p. 171), the kinetics of the behavior of a polar chain molecule in the electric field is determined by relative values of relaxation times of its orientation τ_0 and deformation τ_d. Since τ_0 increases with M proportionally to $M[\eta]$, and τ_d is independent of M, it might be expected that at relatively high M the inequality $\tau_0 > \tau_d$ will hold and, hence, the polarization and the anisotropy of the solution in the electric field will follow the deformational mechanism.

The transition from the orientational to the deformational process with increasing M should be manifested by the decrease of coefficient C in the equation relating the experimental value of τ to $M[\eta]$

$$\tau = CM[\eta]\eta_0/RT \tag{62'}$$

In fact, with the orientation mechanism of EB, C is $1/2$ F. When M increases and the chain conformation changes from rod-like to spherical, C can change correspondingly from C = 3.75 (at F = 2/15) to C = 1.2 (at F = 5/12).

With increasing M the intramolecular motions occurring in a Gaussian chain by various possible modes (Fig. 67) are developed and coefficient C corresponding to these modes can be calculated using the hydrodynamic interactions in the molecule[56].

Thus, for strong hydrodynamic interactions (characteristic of high molecular weight fractions).

$$C_k = 1/0.586 \, \lambda'_k$$

where the values of λ'_k are tabulated for different k (k is the mode number). Applying this equation, the values of the coefficients $C_1 = 0.422$, $C_2 = 0.134$ and $C_3 = 0.071$ are obtained.

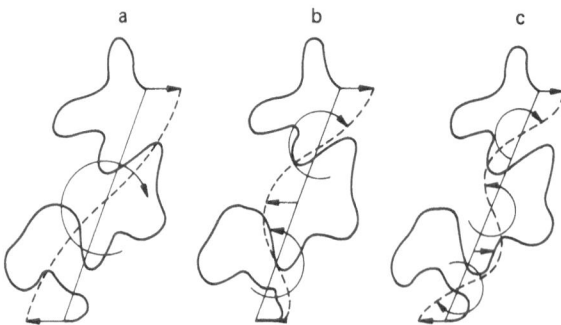

Fig. 67a–c. Various modes of intramolecular motion in the Gaussian chain **a** first mode = rotation of a single-segment chain as a whole: **b** second mode = rotation of parts of a two-segment chain; **c** third mode = rotation of parts of a three-segment chain

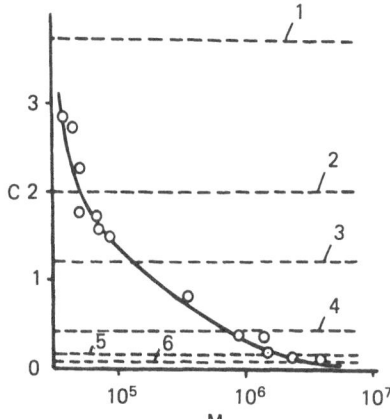

Fig. 68. C vs. M plot of cellulose carbanilate samples in dioxane[215]. Broken lines correspond to the theoretical values of C. 1: rigid rod, 2: rigid coil, 3: rigid sphere, 4: first mode, 5: second mode, 6: third mode

Hence, when passing from the orientational mechanism of EB to the deformational mechanism coefficient C in Eq. (84) decreases and continues to decrease with the introduction of higher modes of intramolecular motion.

Figure 68 illustrates the dependence of C on molecular weight obtained in the investigation of the dispersion of EB for fractions of cellulose carbanilate in dioxane[215]. Coefficient C strongly decreases with increasing M and even at M of $\sim 10^6$ it becomes lower than the minimum theoretical value for kinetically rigid molecules. A further decrease in C with rising M indicates a progressive increase in the contribution of the deformational mechanism to EB.

It might be expected that at relatively high molecular weight the rise in M $[\eta]$ will be compensated by the corresponding lowering of C and, hence, relaxation time τ in Eq. (62') will be independent of M (a kinetically flexible chain comprised of many segments). This case seems to occur with comb-like molecules containing mesogenic side groups (see Sect. 5.9.2 and with high molecular weight deoxyribonucleic acid[225].

5.7 Equilibrium Electro-Optical Properties

5.7.1 Dependence on Field Strength

It has been shown that the main mechanism responsible for EB in solutions of rigid-chain polymers is the rotation of their polar molecules as a whole whereas the anisotropy of the dielectric polarizability of the macromolecules only provides a small contribution to the Kerr effect. Hence, the general theory of the Kerr effect for rigid dipole particles with axial symmetry of the optical polarizability[182, 185, 226] can be used for the description of equilibrium values of EB in these solutions.

According to this theory the value of the excess specific birefringence of the polymer solution $\Delta n/c$ is given by

$$\left(\frac{\Delta n}{c}\right)_{c \to 0} = \frac{2}{9} \frac{\pi N_A}{M} (\gamma_1 - \gamma_2) \cdot \frac{3 \cos^2 \theta - 1}{2} \cdot \frac{(n^2 + 2)^2}{n} \cdot [1 - 3 \mathscr{L}(\varsigma)/\varsigma] \qquad (84)$$

where θ is the angle formed by the dipole moment μ of the molecule with the axis of its optical polarizability (corresponding to γ_1). $\mathscr{L}(\varsigma)$ is the Langevin function of the parameter $\varsigma = \mu \mathscr{E}/kT$. Here $\mathscr{E}$ is the effective electric field related in the Lorenz approximation to the macroscopic field E applied to the solution and described by the Equation $\mathscr{E} = (\epsilon + 2)E/2$ where ϵ is the dielectric permittivity of the solution.

In the range of weak fields, if the first two terms in the series expansion of the Langevin function $\mathscr{L}(\varsigma) = \varsigma/3 - \varsigma^3/45 + 2\varsigma^5/945 - \ldots$ are used and the effect of the internal field is taken into account according to Lorenz, Eq. (84) becomes

$$(\Delta n/cE^2)_{c\to 0,\, \varsigma\to 0} = B_1 (\gamma_1 - \gamma_2)(\mu^2/M)(3\cos^2\theta - 1) \tag{85}$$

where B_1 is expressed by

$$B_1 = \pi N_A (n^2 + 2)^2 (\epsilon + 2)^2/[1215\, n\, (kT)^2] \tag{85'}$$

Equation (85) predicts the proportionality of Δn to the square of the field strength E^2 (the Kerr law). This dependence is experimentally valid at relatively low E for all non-ionogenic rigid-chain polymer solutions if these are fairly pure[210-219] (Fig. 69). The validity of the Kerr law for rigid-chain polymers justifies the introduction of the Kerr constant (determined by use of Eq. (78)) as the main characteristic of EB.

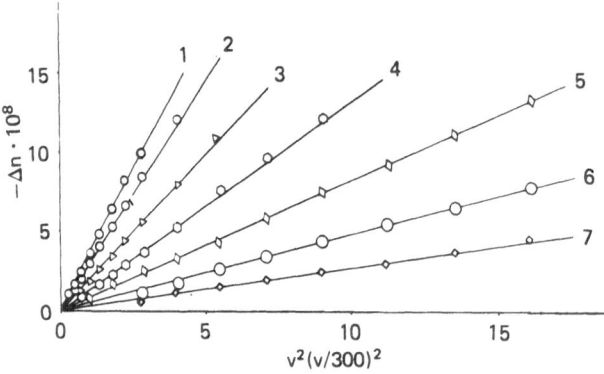

Fig. 69. Plot of birefringence Δn vs. square of the potential difference between the electrodes $V^2 = E^2/d^2$ for a cellulose carbanilate sample in dioxane (M = 8.4 · 10^4)[213, 227] (distance between the electrodes of the cell d = 0.025 cm). c · 10^2 = 0.885 (1); 0.653 (2); 0.407 (3); 0.298 (4); 0.180 (5); 0.107 (6); 0.057 (7) g/cm^3

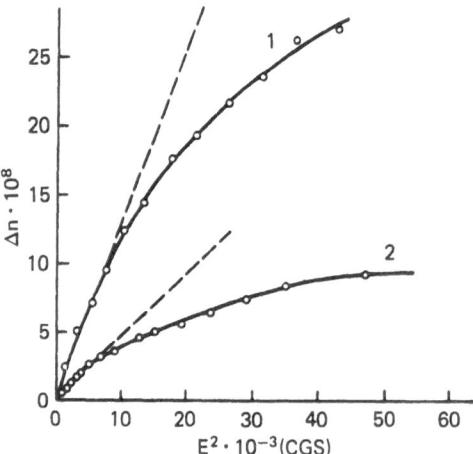

Fig. 70. Graph of EB Δn vs. square of field strength E^2 for solutions of two poly(butyl isocyanate) samples in tetrachloromethane[228]. 1: M = 1.6 · 10^5, c = 0.6 · 10^{-3} g/cm^3; 2: M = 0.66 · 10^5, c = 0.7 · 10^{-3} g/cm^3

At relatively high values of the ζ parameter (i.e. for molecules with relatively high μ), the experimental dependence $\Delta n = f(E^2)$ can the more deviate from linearity the higher E in accordance with the general dependence in Eq. (84) (Fig. 70). The curves in Fig. 70 can be applied to the determination of the dipole moment of the molecule μ. For this purpose, the experimental curves should be compared with the theoretical dependence described by Eq. (84) using the first three terms in the series development of the function $\mathscr{L}(\zeta)$ (up to ζ^5)[147]. However, it should be borne in mind that μ depends on the choice of the factor of the internal field in the $\mathscr{E}$–E relationship.

Equation (85) represents a general relationship between the Kerr constant K and the dipolar and optical properties of a kinetically rigid particle. To establish the quantitative dependence of K on the conformation and structure of a rigid-chain polymer molecule, the molecular model describing its electro-optical properties should be specified. For this purpose, we use a kinetically rigid worm-like chain, just as for the study of the FB problem.

5.7.2 Dipole Moment

Equation (85) shows that the dipole moment of the molecule, μ, is of great importance for EB. For a worm-like chain the square of this dipole moment averaged over all chain conformations $\langle \mu^2 \rangle$ by analogy with the expression (3) (p. 98) is determined from the equation[202]

$$\langle \mu^2 \rangle = N \mu_s^2 [1 - (1 - e^{-x})/x] \tag{86}$$

where N is the number of Kuhn segments in the chain and μ_s the dipole moment of the segment determined as the arithmetical sum of dipole moments of monomer units in the segment (S), $\mu_s = \mu_0 S$. Since $N = M/M_s = M/M_0 S$, Eq. (86) is equivalent to the expression

$$\langle \mu^2 \rangle / M = (\mu_0^2/M_0) S [1 - (1 - e^{-x})/x] \tag{86'}$$

Equation (86') predicts the dependence of $\langle \mu^2 \rangle / M$ on $M = M_0 S x / 2$ coinciding with the dependence of $\langle h^2 \rangle / M$ on M shown in Fig. 1 (Curve 1).

The experimental dependence described by Eq. (86') can be checked if $\langle \mu^2 \rangle$ is ascertained for a number of polymer fractions from measurements of dielectric permittivities ϵ of their dilute solutions by using the well-known Debye equation

$$\langle \mu^2 \rangle / M = (9 \, kT/4 \, \pi \, N_A c) \, [(\epsilon_0 - 1)/(\epsilon_0 + 2) - (\epsilon_\infty - 1)/(\epsilon_\infty + 2)] \tag{87}$$

where ϵ_0 and ϵ_∞ are limiting values of ϵ at frequencies $\nu \longrightarrow 0$ and $\nu \longrightarrow \infty$, respectively.

The experimental dispersion curves which are similar to those in Fig. 66 yields the values of $\langle \mu^2 \rangle / M$ for fractions of poly(chlorohexyl isocyanate) and cellulose carbanilate versus molecular weight M (Fig. 71). The general character of experimental curves corresponds to the theoretical dependence depicted in Eq. (86'). This permits the determination of the dipole moment of the monomer unit, μ_0, from the initial slopes $(\mu_0/M_0)^2$ and the calculation of the parameter of chain rigidity S from their limit $(\mu_0^2/M) \cdot S$.

The values of S obtained in this manner (Table 14) and those obtained by other methods (Tables 1 and 9) are close to each other within experimental error whereas the values of μ_0 (Table 14) and the values that could be expected taking into account the structure of the main chain of these polymers are in reasonable agreement. This means that equilibrium dielectric properties of rigid-chain polymer solutions can be adequately described in terms of the model of a worm-like chain according to Eqs. (86) and (87).

It should be noted that at high M, i.e. in the range of curves approaching a limiting value (Fig. 71) the character of the dependence of $\langle \mu^2 \rangle / M$ on M for rigid-chain polymers and common

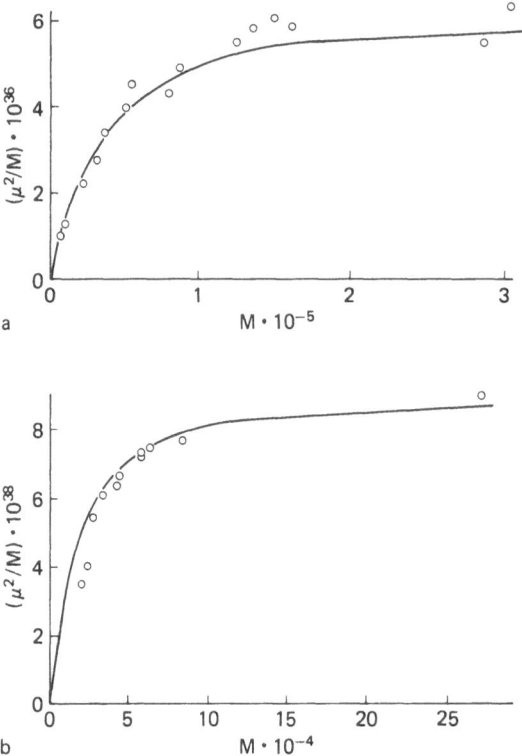

a

b

Fig. 71 a, b. $\langle\mu^2\rangle/M^2$ vs. M plot of poly(chlorohexyl isocyanate) in tetrachloromethane (a)[212] and cellulose carbanilate in dioxane (b)[223]

flexible-chain polymers is the same for any M expect for very low M ($\langle\mu^2\rangle/M$ is constant). However, the use of the limiting values of $\langle\mu^2\rangle/M$ for rigid-chain molecules results in values of S as high as ten and even several hundreds whereas S for flexible-chain molecules is invariably S ⩽ 1. This clearly shows the difference in the mechanisms of polarization of rigid and flexible polymers. For the rigid-chain, a Gaussian coil in "frozen" conformation rotates as a whole and in the flexible chain each monomer unit is virtually independent of others (values of S < 1 usually reflect interactions preventing their rotation[10]).

5.8 EB in an Assembly of Worm-Like Chains

5.8.1 Dipole Direction in a Worm-Like Chain

Equations (85) and (87) clearly show the difference in the equilibrium dielectric and electro-optical properties of rigid-chain polymer solutions. Dielectric polarization depends on the absolute value of the dipole moment of the molecule whereas the value and sign of K are also profoundly affected by the angle θ formed by the molecular dipole μ and the optical axis of the molecule γ_1. In an assembly of statistically bent chain molecules, if Eq. (85) is used, the optical ($\gamma_1 - \gamma_2$) and dipolar μ^2 ($3\cos^2\theta - 1$) factors should be averaged over all conformations.

For a worm-like chain the difference between the main polarizabilities averaged over all conformations $\langle \gamma_1 - \gamma_2 \rangle$ is determined by Eqs. (46) – (48) where γ_1 corresponds to direction h of the end-to-end distance. Hence, the angle θ in Eq. (85) is the angle formed by the direction of the dipole moment μ of a chain molecule and vector h.

If the dipole moment of the monomer unit μ_0 rigidly attached to the chain forms an angle ϑ with the chain direction, it follows that μ_0 is a geometrical sum of the components, parallel $(\mu_{0\parallel} = \mu_0 \cos \vartheta)$ and perpendicular $(\mu_{0\perp} = \mu_0 \sin \vartheta)$ to the chain direction. Hence

$$\mu_0^2 = \mu_{0\parallel}^2 + \mu_{0\perp}^2; \; \mu_{s\parallel} = \mu_s \cos \vartheta; \; \mu_{s\perp} = \mu_s \sin \vartheta; \; \mu_s^2 = \mu_{s\parallel}^2 + \mu_{s\perp}^2 \tag{88}$$

where $\mu_{s\parallel}$ and $\mu_{s\perp}$ are the parallel (longitudinal) and the perpendicular (normal) components of the dipole moment of the segment μ_s, respectively.

The total dipole moment of a chain molecule μ in any conformation is a geometrical sum of dipoles μ_0 of all its monomer units.

It can be seen (Fig. 72a) that in this summation the direction of the part of molecular dipole μ comprised of the sum of longitudinal components of monomer dipoles $(\Sigma \, \mu_0 \cos \vartheta)$ coincides with vector h and, hence, in Eq. (85) we have $\theta = 0$ and $3 \cos^2 \theta - 1 = 2$. Consequently, the dipole factor in Eq. (85) obtained by summing over all $\mu_{0\parallel}$ of a worm-like chain and taking into account Eq. (86') is given by

$$\langle (\mu^2/M) \, (3 \cos^2\theta - 1) \rangle_{\mu_{0\parallel}} = (\mu_{0\parallel}^2/M_0) \, S \, [1 - (1 - e^{-x})/x] \cdot 2 \tag{89}$$

An essentially different result is obtained if normal components of monomer dipoles $\mu_{0\perp}$ rigidly bonded to the main chain are summed. Kuhn has shown[230] that in this case for a molecule described by a chain of freely jointed segments, the direction of μ generally does not coincide with h forming with it angle θ (Fig. 72b) determined by the equation

$$\langle \cos \theta \rangle = (h/L)/\mathscr{L}^* \, (h/L) \tag{90}$$

where $\mathscr{L}^*$ is the inverse Langevin function represented by Eqs. (43a) and (43b).

Equation (90) shows that for a relatively long chain when $h/L \longrightarrow 0$, and, correspondingly, $\mathscr{L}^* \longrightarrow 3 h/L$, the mean square cosine $\langle \cos^2 \theta \rangle \longrightarrow 1/3$. This means that in the Gaussian range the direction of the dipole moment of a chain molecule μ obtained by summation over all $\mu_{0\perp}$ is not correlated with the h direction. Hence, in the Gaussian range, normal components $\mu_{0\perp}$ do not contribute to the Kerr effect (since for these $\langle \cos^2 \theta - 1/3 \rangle = 0$). In this case, the correlation

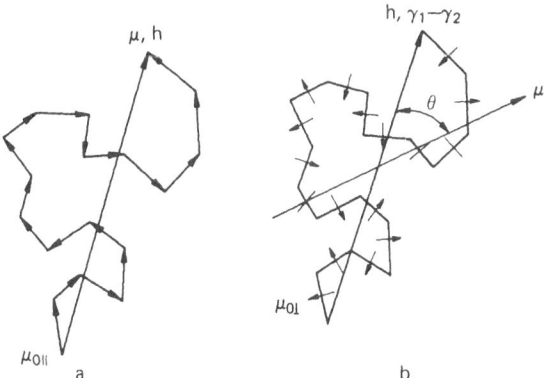

Fig. 72a, b. Direction of the total dipole moment μ in a chain molecule: **a** for the dipole moment of monomer unit $\mu_{0\parallel}$ parallel to the chain; **b** for the dipole moment of monomer unit $\mu_{0\perp}$ normal to the chain

between the μ and h directions appears only with decreasing L (i.e. with increasing h/L) on passing into the non-Gaussian range and becomes complete for the rod-like conformation when h/L = 1, $\mathscr{L} * \longrightarrow \infty$, $\langle \cos^2 \theta \rangle = 0$ and, correspondingly, $\theta = \pi/2$.

Extending this approach from a chain of freely jointed segments to a worm-like chain[231] and applying Eqs. (86') and (90), we find for the dipolar factor in Eq. (85) obtained by summation over all $\mu_{0\perp}$

$$\left\langle \frac{\mu^2}{M} (3 \cos^2 \theta - 1) \right\rangle_{\mu_{0\perp}} = (\mu_{0\perp}^2/M_0) \, S \, [1 - (1 - e^{-x})/x] \, [(3 \, h/L)/\mathscr{L} * - 1] \tag{91}$$

The total value of the dipolar factor in Eq. (85) for a worm-like chain is obtained by the addition of Eqs. (89) and (91). Using Eq. (88) we obtain

$$\langle (\mu^2/M) (3 \cos^2 \theta - 1) \rangle =$$

$$= (\mu_0^2/M_0) \, S \, [1 - (1 - e^{-x})/x] \, [2 \cos^2 \vartheta - \sin^2 \vartheta \, (1 - (3 \, h/L)/\mathscr{L} *)] \tag{92}$$

where substitutions can be made taking into account Eqs. (43b), (45) and (46)

$$1 - (3 \, h/L)/\mathscr{L} * = (3/5) \, (\langle h^2 \rangle/L^2)/[1 - (2/5) \, (\langle h^2 \rangle/L^2)] \tag{93}$$

Equation (92) shows that the dipolar factor is a sum of two terms, the first of which corresponding to the contribution of longitudinal components of monomer dipoles $\mu_{0\parallel}$ is positive whereas the second term representing the contribution of their normal components is negative.

5.8.2 The Kerr Constant

For an assembly of chain molecules of similar length L, the Kerr constant can be obtained by averaging the right-hand side overall chain conformations in Eq. (85) (p. 180)

$$K = B_1 \, \langle (\gamma_1 - \gamma_2) \, (\mu^2/M) \, (3 \cos^2 \theta - 1) \rangle \tag{94}$$

In order to use Eqs. (46) (p. 123), (92) and (93) for this purpose, it should be considered that for a worm-like chain in the Gaussian range where the polydispersity of conformations is most marked, both $\gamma_1 - \gamma_2$ (according to Eq. (46)) and μ^2 (according to Eq. (86'), p. 181) vary proportionally to h^2 with changing conformation. Hence, in complete analogy with Eq. (56), we obtain

$$\langle (\gamma_1 - \gamma_2) \, \mu^2 \, (3 \cos^2 \theta - 1) \rangle/(\langle \gamma_1 - \gamma_2 \rangle \, \langle \mu^2 \, (3 \cos^2 \theta - 1) \rangle) = \langle h^4 \rangle/\langle h^2 \rangle^2 \tag{95}$$

Using Eq. (95), the substitution of Eqs. (46), (92) and (93) into Eq. (94) gives for the Kerr constant

$$K = B_1 \, \frac{3}{5} \, \beta L \, \frac{\mu_0^2}{M_0} \, S \left[1 - \frac{1}{x} \, (1 - e^{-x}) \right] \frac{\langle h^2 \rangle/L^2}{1 - \frac{2}{5} \, \langle h^2 \rangle/L^2} \cdot \frac{\langle h^4 \rangle}{\langle h^2 \rangle^2} \, x$$

$$\times \left[2 \cos^2 \vartheta - \frac{3}{5} \, \sin^2 \vartheta \, \frac{\langle h^2 \rangle/L^2}{1 - \frac{2}{5} \, \langle h^2 \rangle/L^2} \right] \tag{96}$$

If $\langle h^2 \rangle$ is expressed according to Eq. (3) and if it is taken into account that $\beta L = \Delta a \cdot S \cdot x/2$, Eq. (96) becomes

$$K = B_1 \frac{6 \cdot \Delta a \cdot \mu_0^2 S^2}{5 M_0} \cdot \frac{(x - 1 + e^{-x})^2}{x^2 - 0.8(x - 1 + e^{-x})} \cdot \frac{\langle h^4 \rangle}{\langle h^2 \rangle^2} \times$$

$$\times \left[\cos^2 \vartheta - 0.6 \sin^2 \vartheta \frac{x - 1 + e^{-x}}{x^2 - 0.8(x - 1 + e^{-x})} \right] \qquad (97)$$

where B_1 and $\langle h^4 \rangle / \langle h^2 \rangle^2$ are expressed by Eqs. (85') (p. 180) and (57) (p. 126), respectively.

The Kerr constant in Eq. (97) is expressed as a function of the structural parameters Δa, M_0, μ_0, S, ϑ and the reduced length x of the worm-like chain.

5.9 EB, Structure and Conformation of Polymer Chains

5.9.1 Linear Chain Molecules

At low M ($x \longrightarrow 0$) it follows from Eq. (97) that

$$K_{x \longrightarrow 0} = B_1 \cdot P^2 \cdot \Delta a \cdot \mu_0^2 (3 \cos^2 \vartheta - 1)/M_0 \qquad (98)$$

This means that the Kerr constant increases linearly with the square of the degree of polymerization $P^2 = (M/M_0)^2$ as is to be expected for the rod-like molecule in which both the dipole moment and the difference between the main polarizabilities are proportional to the chain length.

In this case, the sign of EB can either coincide with that of the anisotropy of the monomer unit Δa or be opposite to it, depending on the value of the angle ϑ in Eq. (98). Hence, in principle, the signs of EB and FB for a polymer in the rod-like conformation can either coincide or be opposite to each other.

In the Gaussian range when x approaches infinity, the Kerr constant attains the limitting value K_∞ which, according to Eq. (97), is given by

$$K_\infty = 2 B_1 \cdot \Delta a (\mu_0^2 \cos^2 \vartheta) S^2/M_0 \qquad (99)$$

Equation (99) shows that the sign of K_∞ coincides with that of Δa since in this case the Kerr effect is determined by the *square* of the longitudinal component of the monomer dipole $(\mu_0 \cos \vartheta)^2$. Because in the Gaussian range the sign of FB always corresponds to that of Δa (Eqs. (59) and (67)), according to Eq. (99), the signs of EB and FB in a solution of kinetically rigid chain molecules of relatively high molecular weight should coincide. This prediction of the theory has been confirmed experimentally for rigid-chain polymers (Table 13): the characteristic values of $\Delta n/\Delta \tau$ and K widely differing for various rigid-chain polymers always coincide in sign for a solution of the same polymer.

This is a manifestation of the general property of chain molecules oriented in the electric field by the mechanism of large-scale motion: on the average, the three main directions in the molecule coincide: the direction of the greatest geometrical length of the molecule (vector h), of the orientational-axial order (responsible for the anisotropy $\gamma_1 - \gamma_2$) and of the orientational polar order (determining the total dipole moment μ of the molecule).

In Table 13 are also listed, for a comparison, the corresponding experimental data for a typical flexible-chain polymer, polystyrene, and its monomer.

For a styrene monomer both $\Delta n/\Delta \tau$ and K are positive. The positive sign of the former is due to the fact that the direction of the greatest geometrical length of styrene molecules (i.e.

Table 13. Magnitude and sign of the characteristic values of dynamic ($\Delta n/\Delta \tau$) and electric (K) birefringence of some high molecular weight polymers in solution

Polymer	Solvent	$(\Delta n/\Delta \tau) \times 10^{10}$ $(g^{-1} \text{ cm s}^2)$	$K \times 10^{10} \text{ cm}^5 \text{ g}^{-1}$ $(V/300)^{-2}$	Ref.
1	2	3	4	5
Poly(γ-benzyl-L-glutamate)	dichloroethane	+1100	+30000	139)
Poly(butyl isocyanate)	tetrachloro-methane	+300	+25000	137, 148)
Poly(chlorohexyl isocyanate)	tetrachloro-methane	+250	+6400	138, 148)
Ladder poly(phenylsiloxane)	benzene	−160	−12.5	32, 33)
Ladder poly(m-chloro-phenylsiloxane)	benzene	−300	−15	216)
Ladder poly(phenyl-isobutylsiloxane) (1:1)	benzene	−69	−4.9	32, 33)
Ladder poly(phenyl-isohexylsiloxane) (1:1)	benzene	−81	−8.5	32)
Ethyl cellulose	dioxane	+30	+25	218)
Benzyl cellulose	dioxane	+25	+6	218)
Cellulose diphenyl-phosphonocarbamate	dioxane	+50	+25	109, 218)
Cellulose butyrate	dioxane	+4.7	+0.4	218)
Cellulose diphenylacetate	dioxane	+82	+0.9	218)
Cellulose benzoate	dioxane	−35	−10	218)
Cellulose carbanilate	dioxane	−144	−110	125, 215)
Cellulose nitrate; % N = 13.5	ethyl acetate	−80	−40	
%N = 10.7	dioxane	+11	+7	83)
Poly[p-(p-cetylbenzoyloxy)-phenyl methacrylate]	tetrachloro-methane	−220	−8	162)
Monomer: phenyl methacrylic ester of p-cetyloxybenzoic acid	tetrachloro-methane		+0.21	162)
Poly(p-nonyloxybenzamino-carbonylstyrene)	tetrachloro-methane	−210	−140	232)
Poly(cetyl ester of p-methacryloyloxybenzoic acid)	tetrachloro-methane	−30	−0.5	233)
Monomer: cetyl ester of p-methacryloyloxybenzoic acid	tetrachloro-methane		+0.15	233)
Polystyrene	tetrachloromethane	−10	+0.04	64, 169)
Styrene	tetrachloro-methane	+1	+0.07	113, 169)

the direction of the valence bond attaching the phenyl ring to the vinyl group) is close to that of the highest optical polarizability of the molecule (as for all low molecular liquids[6])). The positive sign of K means that the dipole moment μ of the styrene molecule is also close to this direction.

Therefore, the axis of the highest polarizability of the monomer unit of polystyrene is inclined to the axis of the main chain by a wide angle, i.e. the Δa value is negative and, accordingly (see Eq. (67), p. 137), FB is also negative. However, in this case, in contrast to rigid-chain polymers, the Kerr constant is positive, independent of molecular weight and close to the value of K of the monomer.

This behavior is an evidence of the small-scale mechanism of the molecular motion of polystyrene in the electric field: each monomer unit of the polymer chain is oriented virtually independently of other units, just as in the monomeric styrene. Equation (99) shows that for high molecular weight polymers K is proportional to S^2. This accounts for the much higher values of K for rigid-chain polymers than for flexible-chain polymers (Table 13) since the corresponding values of S are of order of magnitudes $10-10^2$ for flexible and approximately unity for rigid polymers in the electric field.

The dependence of the Kerr constant on molecular weight (reduced chain length x) during the corresponding change in the conformation of the molecule from the straight rod to the Gaussian coil is depicted by Eqs. (97) or (100):

$$\frac{K}{K_\infty} = \frac{3}{5} \frac{\langle h^4 \rangle}{\langle h^2 \rangle^2} \cdot \frac{(x - 1 + e^{-x})^2}{x^2 - 0.8\,(x - 1 + e^{-x})} \cdot \left[1 - 0.6\,\mathrm{tg}^2\,\vartheta\,\frac{x - 1 + e^{-x}}{x^2 - 0.8\,(x - 1 + e^{-x})} \right] \tag{100}$$

The character of the dependence of K on x greatly depends on the angle ϑ formed by the monomer dipole μ_0 (rigidly bonded to the chain) and the chain direction. The parallel component of this dipole provides a positive (corresponding to the sign of Δa) contribution to the Kerr effect. It is determined by the first term in Eq. (97) or (100). The normal component of the monomer dipole $\mu_0 \sin \vartheta$ yields a negative (opposite to the sign of Δa) contribution to EB (second term in Eqs. (97) and (100)).

The plots of $K/K_\infty,\, \vartheta = 0$ versus x at various ϑ are shown in Fig. 73. Here we have

$$K_{\infty,\vartheta=0} = K_\infty/\cos^2 \vartheta = 2\,B_1 \cdot \Delta a \cdot \mu_0^2 \cdot S^2/M_0 \tag{101}$$

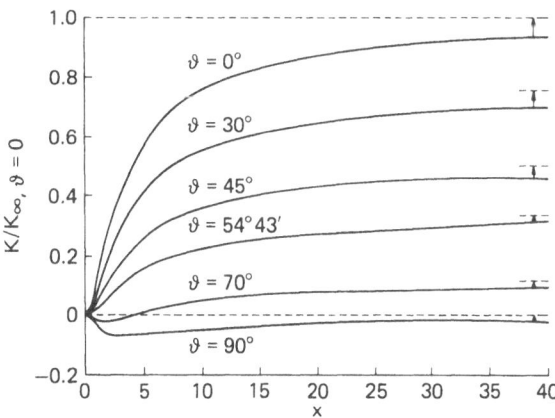

Fig. 73. Graph of relative Kerr constant $K/K_\infty,\, \vartheta = 0$ vs. $x = 2L/A$ for a kinetically rigid wormlike chain. Figures on curves denote angle ϑ formed by the dipole of the monomer unit and the chain direction

The curves in Fig. 73 are distinguished not only by the limiting values (proportional to $\cos^2 \vartheta$) but also by their shape. If the dipole μ_0 is normal to the chain direction ($\vartheta = \pi/2$), the sign of the Kerr effect is opposite to that of Δa at all values of x and when x increases the absolute value of Δa decreases to zero. (Qualitatively, a similar dependence has been observed experimentally in polyoxyethylene oligomers[184]). At any value of ϑ the contribution of the normal component of dipole $\mu_0 \sin \vartheta$ rises with decreasing x and, hence, at $90° > \vartheta > 54,74°$ – according to the theory – the reversion of sign of curves of K versus x is possible. However, this conclusion is related, to a certain extent, to the assumption of the axial symmetry of the optical properties of the monomer unit, the segment and the whole molecule and of the rigidity of the

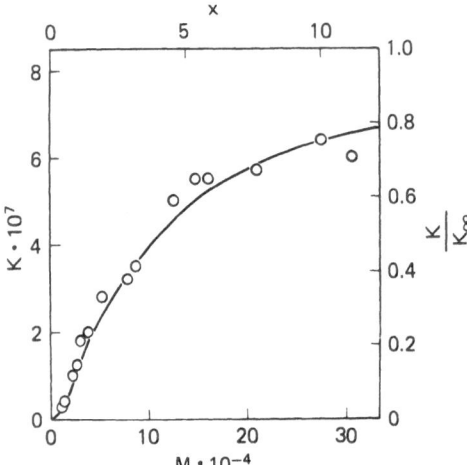

Fig. 74. K vs. M plot for poly(chlorohexyl isocyanate) fractions in tetrachloromethane[210] (points denote experimental data on K vs. M). The curve characterizes the theoretical dependence K/K_∞ vs. x at the values of Δa, S, μ_0 and ϑ listed in Table 14

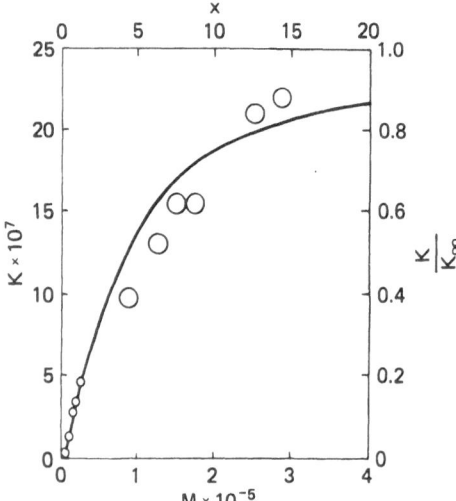

Fig. 75. K vs. M plot for poly(butyl isocyanate) fractions in carbon tetrachloride (points = experimental data[217] on K vs. M). The curve characterizes the theoretical dependence K/K_∞ vs. x at the values of Δa, S, μ_0 and ϑ listed in Table 14

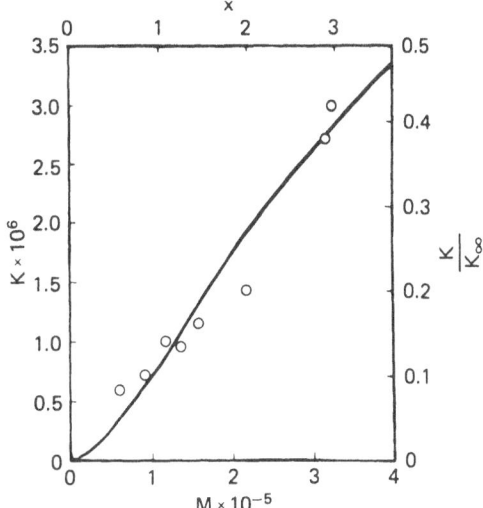

Fig. 76. K vs. molecular weight M plot of poly(γ-benzyl-L-glutamate) fractions in dichloroehane (points = experimental data [202] on K vs. M). The curve describes the theoretical dependence of K/K_∞ on x at the values of Δa, S, μ_0 and ϑ in Table 14

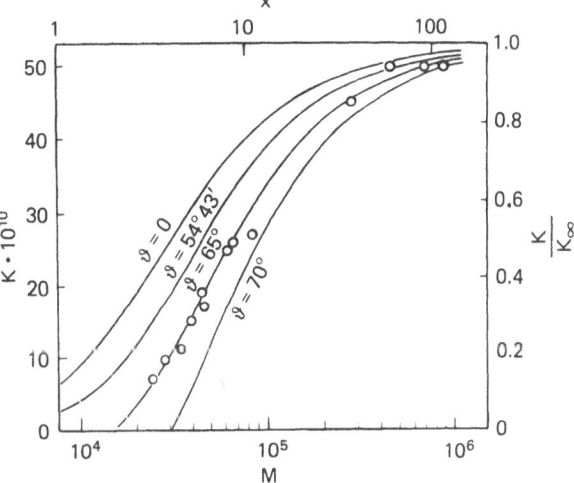

Fig. 77. K vs. M plot of cellulose carbanilate fractions in dioxane (points = experimental data [212] on K vs. M). Curves characterize theoretical dependences of K/K_∞ on x at various values of ϑ shown here and the values Δa, S and μ_0 (Table 14)

attachment of the dipole μ_0 to the polymer chain. A comparison of experimental values of K and of its dependence on M with the data on Δa and S obtained from the FB studies in the same solutions permits the determination of μ_0 and the direction (angle ϑ) of the dipole moment of the monomer unit in the polymer chain. Some examples are given in Figs. 74–77 in which are plotted experimental values of K vs. molecular weight for some rigid-chain polymers.

The experimental points are in satisfactory agreement with the general shape of theoretical curves. The values of μ_0 obtained by a comparison of dynamo-optical with electro-optical data are in reasonable agreement with the values determined by a more direct dielectric method

Table 14. Parameters of rigidity S, optical anisotropy of monomer unit Δa and dipole moments μ_0 of some polymers according to the data of dielectrical polarization measurements and FB and EB of their solutions

Polymer	Solvent	Dielectric measurements		FB data (Tables 6 and 9)		EB data		
		S	$\mu_0 D$	$\Delta a \times 10^{25}$ cm^3	S		ϑ	$\mu_0 D$
Poly(butyl isocyanate)	tetrachloro-methane	500	1.8[229]	15.3	400		0	1.6[217]
Poly(chlorohexyl isocyanate)	tetrachloro-methane	250	1.9[212]	16	340		0	1.3[210]
Cellulose carbanilate	dioxane	50	0.9[223]	−57	35		65	2.3[212]
Poly(γ-benzyl-L-glutamate)	dichloro-ethane	−	−	34	1000		0	0.4[202]

(Table 14). The differences in the μ_0 values obtained by these two methods may be caused by some uncertainty of the internal field factor in B_1 (Eq. (85'), p. 181). The experimental data on poly(alkyl isocyanate)s and poly(γ-benzyl-L-glutamate) yield $\vartheta = 0$ which is in good agreement with the axially symmetric structure of the main chain of these polymers. It is shown in Fig. 77 that for cellulose carbanilate the agreement between the theoretical curve and experimental points is attained at $\vartheta = 65°$. Evidently, in this case the side group of the monomer unit containing polar bonds provides an important contribution to the dipole moment of the monomer unit.

5.9.2 Comb-Like Polymers

Many polymers with comb-like molecules are typical flexible-chain polymers as revealed by their hydrodynamic properties and FB in their solutions (Sect. 4.6).

Their electro-optical properties are also in agreement with the following finding: the Kerr constant in solutions of these polymers is of the order of 10^{-12} cm^5 g^{-1} $(V/300)^{-2}$; it is independent of molecular weight and its sign does not correlate with that of FB.

The electro-optical properties of comb-like molecules with mesogenic side groups are unique and differ from those of other comb-like molecules just as FB in solutions.

Apart from poly(phenyl methacrylic ester)s of p-alkoxybenzoic acids (PPhEAA) (p. 165), molecules of this type include e.g. poly(p-nonyloxybenzamidostyrene) (PNO BS)[232]

and poly(cetyl ester of p-methacryloyloxybenzoic acid) (PCEMA)[233].

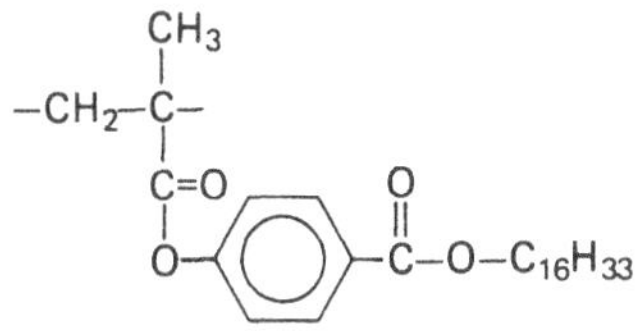

CH3

—CH2—C—

    C=O

    O — ⬡ — $\overset{O}{\overset{\|}{C}}$ —O—C16H33

Shear optical coefficients $\Delta n/\Delta\tau$ and Kerr constants for these polymers (and for PPhEAA) are listed in Table 13. Their K values are high and negative and the signs of K and FB are the same. However, the Kerr constants in solutions of the CEMA, NOBS and PhEAA monomers are positive, just as for styrene, in accordance with the fact that the directions of the dipole moments of these monomers are close to those of their longitudinal axes $L_{\|}$, i.e. to the axes of their maximum optical polarizability ($\mu_{\|} > \mu_{\perp}$, Fig. 78a).

The negative sign of K in polymer solutions of PCEMA, PNOBS and PPhEAA means that the side groups of these comb-like polymers are oriented in the electric field with their long axes normal to the field (in contrast to the corresponding monomer molecules) and, hence, this orientation is not free (as for polystyrene) but is in correlation with that of the main chain.

This can be explained by the fact that in a polymer molecule (Fig. 78b) the longitudinal components of monomer unit dipoles $\mu_{\|}$ are mutually compensated and the main part in the observed EB is played by normal components of monomer unit dipoles, $\mu_{\perp}$, which can be parallel to the main chain of the macromolecule owing to its comb-like structure. In other words, in molecules of comb-like polymers containing mesogenic side chains, the orientations of the $\mu_{\perp}$ components of the side group dipoles are correlated with each other. As a result, the macromolecule as a whole or part of it can exhibit a considerable dipole moment μ in the direction of the main chain L (Fig. 78b). The existence of this dipole accounts for the orientation of the main chain in the field direction leading to negative EB.

Hence, although the equilibrium rigidity of the main chain of polymer molecules with mesogenic side groups is not high (A ≈ 50 A), their polar groups are oriented in the electric field by large-scale intramolecular motions characteristic of rigid-chain polymers.

This conclusion is supported by the existence of low frequency dispersion of EB in solutions of these polymers (Fig. 79)[234]. For all PCEMA fractions studied in the frequency range 0.5 to

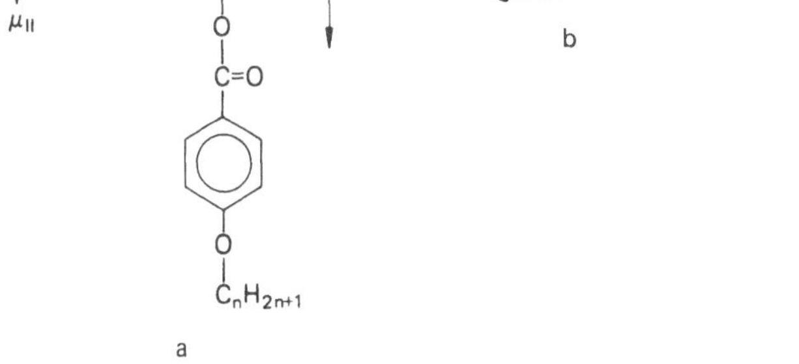

Fig. 78a,b. Explanation of the sign of EB for a monomer (a) and a polymer (b) in solutions of comb-like macromolecules with mesogenic side groups

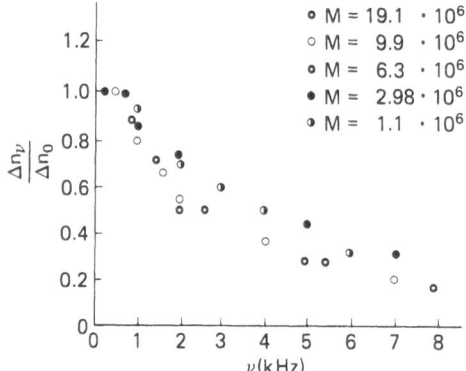

Fig. 79. Relative EB $\Delta n_\nu / \Delta n_0$ as a function of the frequency of the electric field ν for PCEMA fractions in tetrachloromethane with different molecular weights M[163, 234]: $M \cdot 10^{-6} = 19.1$ (1); 9.9 (2); 6.3 (3); 3.0 (4); 1.1 (5)

10 (kHz) the ratio $\Delta n_\nu / \Delta n_0$ decreases sharply which corresponds to relaxation times τ of 10^{-4} to 10^{-5} s. However, for all fractions the values of τ virtually coincide within experimental error. This means that the values of the "kinetic unit", oriented by the action of the electric field, is independent of molecular weight. However, the rotational mobility of this kinetic unit is lower by five to six orders of mangitude than for common flexible-chain polymers. Probably, specific interactions between mesogenic side groups of the comb-like molecule lead to the formation of kinetic segments in this molecule. Each of these segments comprises a large portion of the main chain and the corresponding number of monomer units. Consequently, the whole molecule displays the properties of a kinetically flexible chain with the local rigidity determined by the size of kinetic segments responsible for the intramolecular large-scale motion.

5.10 Electro-Dynamic Birefringence

The data reported in this section demonstrate the advantage of investigating rigid-chain polymers by combining FB and EB because the combination of the results of these two methods provides complete information on the structural and electro-optical characteristics of these molecules.

When this combination of methods is used, it is very important to carry out FB and EB studies of the same polymer at the same concentration and temperature, with the same optical apparatus etc. For this purpose, the method of electro-dynamic birefringence (EDB) offers some advantages. In this method the solution under study passes into the state of a laminar flow and is simultaneously exposed to the electric field E parallel to the velocity gradient g[235-239].

In all cases of practical importance, the anisotropy of solutions resulting from the combined action of dynamic and electric fields is not high: the phase difference of two interfering polarized beams does not exceed $2\pi \cdot 10^{-2}$. Under these conditions, the total birefringence Δn of the solution and the corresponding orientation angle α can be calculated according to[298]

$$\Delta n^2 = \Delta n_g^2 + \Delta n_E^2 - 2 \Delta n_g \Delta n_E \cos 2\alpha_g \tag{102}$$

$$\operatorname{ctg} 2\alpha = \operatorname{ctg} 2\alpha_g - (\Delta n_E / \Delta n_g) \cdot \sin 2\alpha_g \tag{103}$$

where Δn_g and α_g are FB and orientation angle of the optical axis in the absence of the electric field, respectively and Δn_E is EB in the absence of the flow.

Substitution of the expressions for Δn_g, α_g and Δn_E from FB[4] and EB[182] theories of solutions of rigid molecules at low values of g/D_r and $m\mathscr{E}/kT$ into Eqs. (102) and (103) gives

$$\left(\frac{\Delta n}{g \eta_0 c}\right) = \frac{2\pi N_A}{135 M \eta_0 kT} \frac{(n^2 + 2)^2}{n} bW (\gamma_1 - \gamma_2) \left[1 + \left(\frac{m\mathscr{E}}{kT}\right)^4 \left(\frac{D_r}{bg}\right)^2\right]^{\frac{1}{2}} \qquad (104)$$

$$\text{ctg } 2\alpha = \frac{g}{6 D_r} - \left(\frac{m\mathscr{E}}{kT}\right)^2 \frac{D_r}{bg} \qquad (105)$$

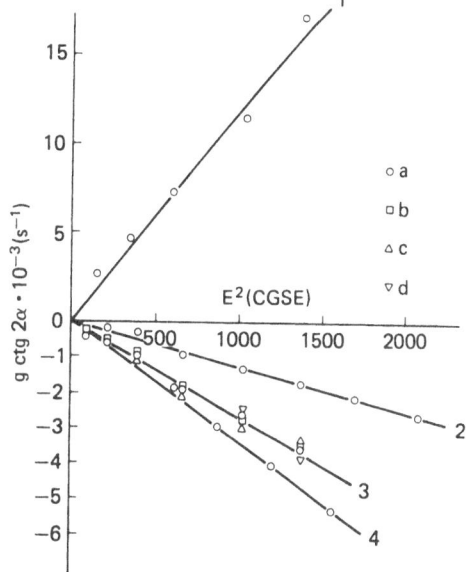

Fig. 80. Orientations of electrodynamic birefringence (EDB) of liquids vs. applied electric field E[238]. 1: Bromoform, g = 3660 s^{-1}; 2: α-methylnaphthalene, g = 1380 s^{-1}; 3: α-bromonaphthalene: a) g = 1380 s^{-1}, b) g = 1180 s^{-1}; c) g = 1035 s^{-1}, d) g = 690 s^{-1}; 4: mixture of α-bromonaphthalene with decalin, g = 1180 s^{-1}

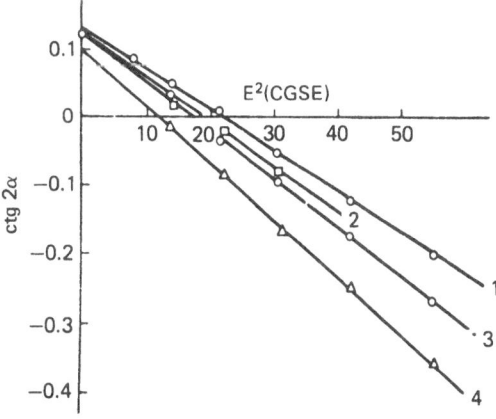

Fig. 81. Orientation of EDB vs. E for solutions of PBLG (M = 1.36 · 10^5) in chloroform[239]; c (in g cm^{-3} · 10^4): 9.1 (1), 7.2 (2), 6.7 (3), 2.3 (4); g = 862 s^{-1}

with

$$m^2 = (b_1 - b_2)kT + \frac{\mu^2}{2} \ (3 \cos^2 \theta - 1)$$

$b_1 - b_2$ is the difference between the two main dielectric polarizabilities of the molecule and $\mathscr{E}$ is the field strength to which the molecule is exposed. (For other symbols see Eqs. (55) (p. 125) and (61), (p. 128).)

Equations (104) and (105) differ from Eqs. (55) and (61), respectively by the presence of the term $(m \ \mathscr{E}/kT)^2 \ (D_r/bg)$ that takes into accound the change in the value and orientation of birefringence caused by the effect of the electric field $\mathscr{E}$.

The validity of Eqs. (102) – (105) for both low molecular weight liquids[238] and rigid-chain polymer solutions[239] was confirmed by appropriate experiments. Examples are given in Figs. 80 and 81 which reveal the dependences of ctg 2 α on $E^2 = \mathscr{E}^2 \ (3/(\epsilon + 2))^2$ at constant g. In accordance with Eqs. (109) and (105) the points fit straight lines the intercepts of which yield ctg 2 α_g and the slope can be used for the determination of m.

The EDB method provides information on molecular hydrodynamic, electric and optical characteristics of the polymer investigated. In principle, these results coincide with those of separate investigations of FB and EB but the fact that the data are obtained in a single experiment offers advantages.

6 Experimental Procedures

The instruments used in the investigations of both flow[64–72] and electric[167, 241, 270] birefringence of solutions, have been widely described.

Hence, we will only briefly discuss the methods of the investigation of solutions of rigid-chain polymers described in previous chapters.

In the study of hydrodynamic characteristics of chain molecules sedimentation and diffusion measurements are of great importance. For a reliable determination of diffusion D and sedimentation [s] coefficients, measurements should be carried out in relatively dilute solutions. Thus, a polarizing interferometer can be used as the optical part of a diffusometer or an ultracentrifuge[6, 52, 53, 249–251].

As already indicated (Sect. 2.2), many rigid-chain polymers are insoluble in organic solvents so that their molecular characteristics can only be studied by use of very "agressive" solvents such as concentrated sulfuric acid. However, sedimentation measurements are not employed in practice and only diffusion studies can provide information about the translational mobility of these polymer molecules in solution.

For this reason, a special diffusion cell was constructed[50] for diffusion measurements in sulfuric acid as the solvent. The parts of cells that are in contact with the solution are made of Teflon and glass. Such cells have been applied successfully to the determination of the translational mobility and, according to Eq. (11), of the molecular weight of aromatic polymers insoluble in organic solvents.

The instrument for the investigation of FB comprises a mechanical part and an optical part.

6.1 Mechanical Part

To achieve a state of laminar flow of the solution to be studied, a cylindrical apparatus (dynamo-optimeter) with an inner or outer rotor is employed.

Figure 82 schematically describes the instrument with an internal rotor widely used for the study of various polymers. Rotor R rotates on two ball bearings P mounted in the upper part of the top which can be taken off with the rotor from stator C. Viewing windows S of thin quartz glass (without tension) allow to observe the operation through gap ΔR in the direction of the axis of the instrument. A thermostating water jacket is placed in the stator (pipes K) and the temperature is controlled with a thermocouple (by opening T). H is the inlet opening for pouring the solution and M the pulley. The width of the rotor-to-stator gap is 0.2 to 0.5 mm, the rotor height is 30 to 50 mm and the rotor diameter 30 to 40 mm. All the details of the instrument coming in contact with the solution are made of titanium and it can therefore be used with any organic solvent and with aqueous polyelectrolyte solutions. The capacity of the instrument is not high: it is fed with only 3 to 5 ml of the polymer solution and may readily be used if several polymer fractions have to be investigated and the amount of each fraction is limited.

In cylindrical instruments with an outer rotor high velocity gradients can be attained with relatively wide gaps without passing the limits of a laminar flow[242].

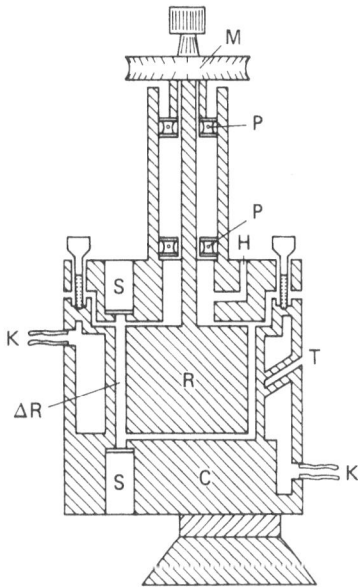

Fig. 82. Cylindrical apparatus with an inner rotor and bearings at the top of the stator

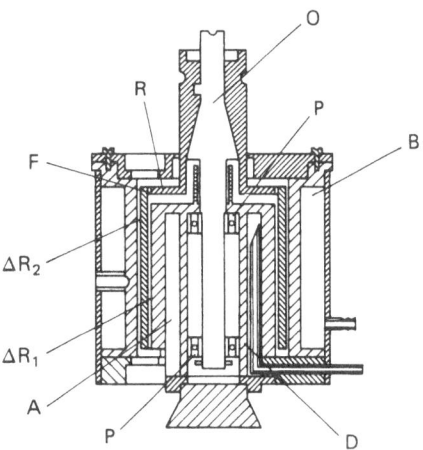

Fig. 83. Cylindrical apparatus with an outer rotor and bearings in the body of the stator

However, this can be actually used to advantage only in instruments with fixed windows[52, 243–245]. One type of this instrument[245] is described in Fig. 83. The rotor of the dynamo-optimeter is a hollow cylinder R open from below and rotating together with the O axis on ball bearings P mounted in the stator D. Inspection through the gap ΔR_1 is carried out with a system of ring-shaped perforations F in the upper base of the rotor. When the inspection is through the gap ΔR_2, the dynamo-optimeter can function as an instrument with an inner rotor. All details in contact with the solutions are made of titanium. The solution is thermostated by using an internal (A) and external system (B). The rotor height usually varies from 50 to 100 mm, the gap is 0.4 to 1.0 mm and the rotor diameter 50 to 80 mm. The capacity of the dynamo-optimeter with the outor rotor is higher by one order of magnitude than that of the instrument with the inner rotor (Fig. 82). Consequently, these instruments are suitable for precise measurements of a weak FB in low viscosity solvents.

For operations with agressive solvents a teflon dynamo-optimeter should be used (Fig. 84). This dynamo-optimeter is equipped with an inner rotor and differs from that described in Fig. 82 in that it contains no metal ball bearings. The step bearing L of teflon rotor R is supported by stator C. The second bearing P is formed directly by the top of the stator. The only metal (titanium) detail of the instrument is the axis O mounted inside the body of the rotor. Optimum dimensions of this instrument are as follows: rotor height 30 to 60 mm, rotor diameter 30 to 50 mm and the gap is 0.4 to 0.8 mm.

The rotor of the cylindrical dynamo-optimeter is activated by a motor through a pulley and a system of reduction gears.

The rotational velocity of the rotor is measured with the aid of a perforated disc that periodically interrupts a light beam during the rotation. The pulsing light beam is fed to a photodiode the electric impulses of which are fed to a digital frequency meter. The frequency of impulses

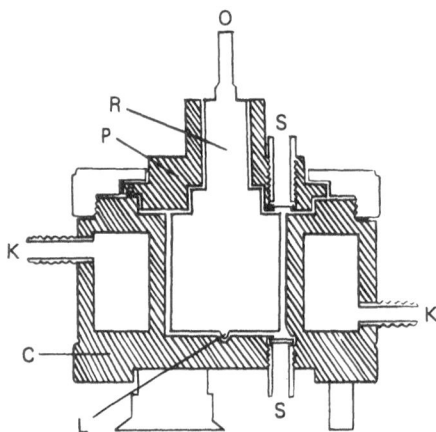

Fig. 84. Teflon cylindrical apparatus for operations with agressive solvents

shown by the meter is proportional to the rotor speed. This scheme permits a reliable measurement of the rotation frequency of the rotor in the range from $2 \cdot 10^{-2}$ to 50 rpm.

In the study of electrodynamic birefringence, metal dynamo-optimeters with both inner and outer rotors can be used. They differ from the constructions shown in Figs. 82 and 83 in that the axis on which the rotor rotates is isolated from the stator with a plastic packing[238]. The electric potential is supplied to the rotor.

6.2 Optical Part

The optical method used in the investigation of EB may differ, depending on the electric conductivity of the liquids under study. For solutions displaying electric

conductivity (polyelectrolytes) the pulsed procedure involving oscilloscopic recording of light pulses should be used[167, 241]. When low conductivity liquids are investigated, compensation methods may be applied. Their sensitivity (value of the measured
birefringence) greatly exceeds that of the pulsed method. Since the electric conductivity of well purified rigid-chain polymer solutions was relatively low, the EB of these
polymers has been studied in sinusoidal fields using a compensation scheme with optical compensators. The same optical scheme has been utilized in the investigation
of the FB of solutions.

In compensation schemes the recording of FB can be conducted either by visual
observation or by use of a photoelectric receiver. It has been found that the optical
anisotropy of rigid-chain polymer molecules is fairly high. However, when very dilute
solutions are studied, the measured excess birefringence can be very low. Hence, to
obtain reliable information on the molecular characteristics of polymers, methods
ensuring maximum sensitivity of the instrument should be used.

6.2.1 Visual Method

Figure 85 schematically describes the optical part of the instrument for the visual study of FB
and EB. A compensating device according to Brace[246] is used in this scheme. It consists of a
thin mica plate K (elliptical compensator introducing phase difference δ_K corresponding to a
few hundredths of a wavelength) rotating on a limb B_2 and a very thin half-shaded plate N (a
few thousandths of wavelength) closing half of the field of vision. Light source S (mercury lamp
of superhigh pressure) is projected on the entrance slit of monochromator M. The image of the
exit slit of the monochromator is projected by lens O on the edge of half-shaded plate N. Crossed
polarizing prisms P and A rotate together with plate N and compensator K with the aid of lever R
fixed on limb B_1. A thin lens L projects the image of plate N into gap D between the stator and

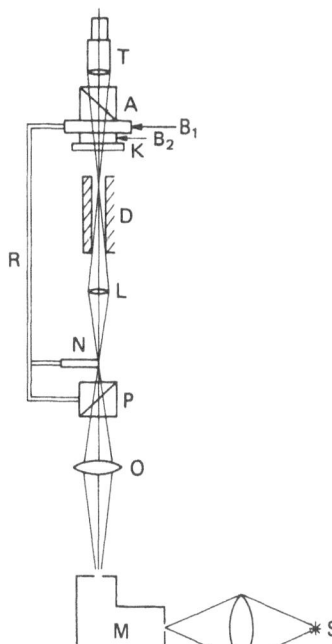

Fig. 85. Scheme of the optical part of the apparatus
for FB and EB with visual observation

the rotor of the cylindrical instrument (in FB measurements) or into the gap between the electrodes of the Kerr cell (in EB measurements). Telescope T is also foccused on this gap.

The principle of the method using the elliptical compensator is based on Eq. (106) expressing intensity I of the light passing through crossed polarizing prisms P and A and two thin anisotropic plates placed between these prisms at azimuths η and η_K with respect to the polarizer P and introducing the phase differences δ and δ_K, respectively[247]

$$I = (I_0/4) (\delta \cdot \sin 2\,\eta + \delta_K \cdot \sin 2\,\eta_K)^2 \tag{106}$$

where I_0 is the intensity of the light emerging from polarizer P.

Equation (106) is approximate and holds for low δ values if $\delta \approx \sin \delta$.

If the parameters δ and η in Fig. 85 refer to the anisotropic layer being investigated, δ_K and η_K refer to the elliptical compensator K, and if the half-shaded plate N is absent, then it follows that according to Eq. (106) the darkening of the field of vision (I = 0) can occur at $\delta \sin 2\,\eta = -\delta_K \sin 2\,\eta_K$. In this case and if the optical axis of the anisotropic layer is in the diagonal position ($\eta = \pi/4$), which occurs in the measurements of δ, the condition of darkening is expressed as $\delta = -\delta_K \sin 2\,\eta_K$ and can be used for the determination of δ from the known value of δ_K by rotating the compensator K (limb B_2) to the position of darkening η_K. Actually, the darkening is incomplete but a very low minimum of the illumination of the field of vision is observed. The precision of mounting the compensator into the compensating position greatly increase if half-shaded plate N is introduced. In this case the desired phase difference δ is determined according to

$$\delta = -\delta_K (\sin 2\,\eta_K - \sin 2\,\eta_0) \tag{107}$$

where η_K and η_0 are half-shaded azimuths of the compensator (i.e. azimuths corresponding to equal illumination of the two halves of the field of vision) registered from the main plane of the polarizer P in the presence and absence of birefringence in solution, respectively. The value of birefringence Δn is determined from that of δ according to $\Delta n = \delta \lambda / 2\,\pi l$ where l is the thickness of the anisotropic layer in the path of light beams and λ the wavelength.

The main optical directions of the anisotropic layer (and, correspondingly, the orientation angle in FB or EB) are found when the compensator is switched on ($\delta_K = 0$ or $\eta_K = 0$ in Eq. (106)) and the half-shaded position is established by rotating the whole optical system (limb B_1). The orientation angle α is determined as the half-angle between two half-shaded positions one of which corresponds to the clockwise rotation of the rotor of the dynamo-optimeter and the other to the counter-clockwise rotation.

6.2.2 Photoelectric Method

The photoelectric method[248] successfully employed in the study of FB and EB also uses the compensation principle based on Eq. (106). However, the anisotropy in solution and its compensation are recorded with the aid of the scheme shown in Fig. 86 rather than visually.

In contrast to the scheme in Fig. 85, in the optical part of this scheme, the half-shaded plate N is replaced by a harmonic modulator 5 of the ellipticity of the polarized light. The main detail of the modulator rigidly secured to the lever 16 is a ferrite shaft in which an alternating magnetic field excites longitudinal mechanical vibrations at the resonance frequency ω of the shaft. Harmonic optical anisotropy (photoelastic effect) is established in a glass plate rigidly fixed to the end of the shaft $\delta_1 = \delta_{10} \sin \omega t$. The axis of this anisotropy forms an angle $\pi/4$ with the axis of analyzer 7.

In this scheme, intensity I of light passing through crossed polarizer 3 and analyzer 7 is determined by the combined action of the anisotropic layer (introducing the phase difference $\delta \cdot \sin 2\,\eta$), the compensator ($\delta_K \sin 2\,\eta_K$) and the modulator ($\delta_1 = \delta_{10} \sin \omega t$). According to Eq. (106) I is given by

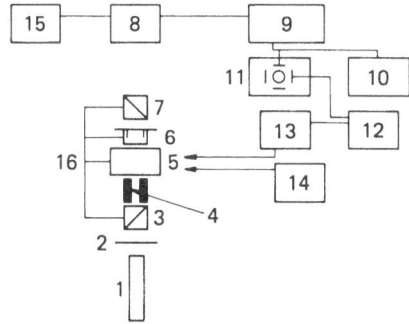

Fig. 86. Scheme of the photoelectric method for birefringence measurements. 1: He-Ne laser ($\lambda = 6300$ A), 2: plate $\lambda/4$ ensuring circular polarization of the beam entering polarizer 3, 4: solvent studied, 5: modulator, 6: compensator, 7: analyzer, 8: photomultiplier, 9: selective (narrow field) amplifier, 10: synchronous detector, 11: oscillograph, 12: phase shifter, 13: sonic generator, 14: modulator power supply, 15: photomultiplier source of power, 16: rotating lever fixing together the mechanical elements 3, 5, 6 and 7

$$I = C \left[\delta \cdot \sin 2\,\eta + \delta_K \cdot \sin 2\,\eta_K + \delta_{10} \sin \omega t \right]^2 =$$

$$= C \left[(\delta \cdot \sin 2\,\eta + \delta_K \cdot \sin 2\,\eta_K)^2 + \frac{\delta_{10}^2}{2} \cos \omega t + \right. \tag{108}$$

$$\left. + 2\,\delta_{10} \,(\delta \sin 2\,\eta + \delta_K \sin 2\,\eta_K) \sin \omega t \right]$$

where C is a constant.

Hence, the intensity of the light emerging from analyzer 7 and reaching photomultiplier 8 is expanded into three components: a constant component and two sinusoidal components with ω and $2\,\omega$ frequencies. The corresponding electric signal at the output of the multiplier is received by a narrow-band resonance amplifier synchronous with modulator 5 and separating from the overall signal a harmonic part at frequency ω. As a result, according to Eq. (108), amplitude V of the harmonic signal fed to oscillograph 11 or to synchronous detector 10 is given by

$$V = 2\,\delta_{10} \,(\delta \sin 2\,\eta + \delta_K \sin 2\,\eta_K) \tag{109}$$

V depends linearly on the anisotropy $\Delta n = \delta \lambda/2\,\pi l$ of the solution. This fact ensures a higher (not less than by an order of magnitude) sensitivity of the photoelectric method than that of the visual method in which, according to Eq. (106), the light flux affecting the eye is proportional to δ^2.

Equation (109) demonstrates that the value of the signal V recorded with an oscillograph or a detector becomes zero at $\delta \sin 2\,\eta = -\delta_K \sin 2\,\eta_K$, i.e. under the same conditions as in visual observation. Hence, the procedures for the determination of the anisotropy of the solution and the orientation angle in the photoelectric and visual methods are identical.

7 Conclusions

As already indicated (p. 98), the interest in rigid-chain polymers in caused to a great extent by the prospects of their practical use. These prospects depend mainly on the possibility of attaining the mesomorphic state in concentrated solutions of these polymers[107, 108]. For polymers containing mesogenic segments, as well as for low molecular liquid crystals[252], this property directly depends on the degree of extention and rigidity of the polymer chain. However, in contrast to low molecular compounds of chain polymers the ability of forming mesophases is governed by the extent of coiling of the chain rather than by its total length, i.e. it is determined by the persistent length a or the length of the Kuhn segment A. Actually, this ability has been established only for those polymers considered before for which A is 400 to 600 A. These polymers include: poly (γ-benzyl-L-glutamate) in the helical conformation[253, 254], poly(alkyl isocyanate)s[255, 256] and *para*-aromatic polyamides[107, 108]. In contrast, for poly(amide hydrazide) the rigidity of which is only slightly lower than that of poly(*p*-phenylene terephthalamide) (Table 9), the mesomorphic state in concentrated solutions has not been observed[108]. These facts demonstrate that the increase in chain flexibility causes a sharp decrease in the capacity of forming mesophases. Hence, the experimental determination of the persistent length of the chain provides important information for the prediction of the mesogenic properties of the polymer.

The material considered in the preceding sections shows that the FB and EB methods can be successfully applied to the solution of these problems providing information not only on the equilibrium but also on the kinetic properties of polymer chains. This is particularly valuable in those cases in which the application of other more common methods is difficult or even impossible (e.g. sedimentation measurements of polymer solutions in sulfuric acid).

On the other hand, the peculiarities of the molecular structure of rigid-chain polymers favor the use of EB whereas the application of this method to flexible-chain polymers is much less effective. To a certain extent, this also refers to FB which permits the determination of both the anisotropy and the rigidity of the polymer chain using the molecular weight dependence of the shear optical coefficient when chain rigidity is relatively high.

The material considered in this review show that the molecules of rigid-chain polymers exhibit a number of characteristic (sometimes even unique) properties many of which cannot be observed in flexible-chain polymers. The FB and EB of rigid-chain polymer solutions can serve as effective methods of studying these properties.

8 List of Symbols

a	persistent length	b	effective chain anisotropy in a system of axes of the first element
$a_{\parallel} - a_{\perp} = \Delta a$	difference between the main polarizabilities of the monomer unit		
		c	solution concentration
b	parameter of the shape asymmetry of the molecule	d	hydrodynamic diameter of the chain
		f	translatory friction coefficient of the molecule

g	velocity gradient	M_n, M_w	number average and weight average molecular weight of the macromolecule
h	end-to-end distance of the chain		
$\langle h^2 \rangle$	mean-(overall conformations) square end-to-end distance of the chain		
		M_0	molecular weight of a monomer unit
k	Boltzmann's constant	M_s	molecular weight of a segment
n	refractice index of the solution	N	number of segments in a chain
$[n]$	characteristic value of FB		
Δn	excess birefringence of the solution	N_A	Avogadro's number
		P	function of hydrodynamic interactions in translatory frictions
Δn_ν	EB of solution at field frequency ν		
Δn_0	EB in a steady field	R	gas constant
dn/dc	refractive index increment in a polymer-solvent system	$\langle R^2 \rangle$	mean-square radius of gyration
		S	number of monomer units in a statistical Kuhn segment
$\Delta n/\Delta \tau$	shear optical coefficient		
p	axial ratio of the ellipsoid		
$p(x)$	shape asymmetry of a worm-like chain	S_f	number of monomer units in a statistical Kuhn segment assuming free rotation about valence bonds
$[s]$	sedimentation coefficient		
$\bar{v}$	partial specific volume of the polymer		
$x = L/a$	relative length of the molecular chain	T	absolute temperature
		W	rotatory friction coefficient of the macromolecule
A	length of the statistical Kuhn segment		
A_f	length of the statistical Kuhn segment on the assumption of free rotation about valence bonds	Z	relative content of a component in the copolymer
A_0	hydrodynamic parameter	α, β	exponents in Mark-Kuhn's equations
B	optical coefficient	α	orientation angle of flow birefringence
C	coefficient	$\alpha_1 - \alpha_2$	optical anisotropy of a segment
D	translational diffusion coefficient		
D_r	rotatory diffusion coefficient	β	optical anisotropy of the unit length of a worm-like chain
E	electric field strength		
$\mathcal{E}$	electric field affecting the particle (molecule)	β_M	optical anisotropy of a chain per unit of its molecular weight
F	coefficient depending on the asymmetry of a worm-like chain	$\gamma_1, \gamma_2, \gamma_3$	main optical polarizabilities of the macromolecule
G	coefficient	ϵ	dielectric permittivity of a solution
K	Kerr's constant	ϵ_0	excess value of dielectric permittivity of a solution in a steady electric field (at $\nu \longrightarrow 0$)
K_η, K_D	coefficients in Mark-Kuhn's equations		
L	contour length of the chain		
M	molecular weight of the macromolecule	ϵ_∞	excess value of dielectric permittivity of the solution at field frequency $\nu \longrightarrow \infty$

$\dfrac{\Delta\epsilon}{\Delta c}$	dielectric increment	μ_s	dipole moment of a segment
ζ	mean effective friction coefficient of a bead	$\mu_{s\parallel},\ \mu_{s\perp}$	parallel (longitudinal) and perpendicular (normal) components of the dipole moment of a segment
ζ	parameter in the Langevin function		
η	viscosity of solution		
η_0	viscosity of solvent	ν	electric field frequency
$[\eta]$	intrinsic viscosity	ρ	solvent density
ϑ	angle formed by the dipole moment of a monomer unit with the chain direction	τ_d	time of deformation of the macromolecule
		τ_0	time of reorientation of the macromolecule as a whole
λ	projection of the monomer unit on the chain direction		
		$\Delta\tau = g\,(n - \eta_0)$	effective shear stress in solution
μ	dipole moment of the chain	$[\chi/g]$	characteristic orientation angle
μ_0	dipole moment of a polar group (monomer unit)	ω	frequency of the sinusoidal electric field (in radians)
		Φ	function of hydrodynamic interactions in viscosity phenomenon
$\mu_{0\parallel},\ \mu_{0\perp}$	parallel and perpendicular components of the dipole moment of the monomer unit with respect to the chain direction	θ	angle formed by the dipole moment of the molecule and the axis of its optical polarizability

9 References

1. Kuhn, W.: Kolloid Z. *68*, 2 (1934)
2. Kuhn, W., Kuhn, H., Buhner, P.: Ergeb. Exakt. Naturwiss. *25*, 1 (1951)
3. Porod, G.: Monatsh. Chem. *80*, 251 (1949)
4. Stuart, H. A. (ed.): Die Physik der Hochpolymeren, Bd. 1 u. 2. Berlin: Springer Verlag 1952, 1953
5. Flory, P. J.: Principles of polymer chemistry. Ithaca, New York: Cornell University Press 1953
6. Tsvetkov, V. N., Eskin, V. E., Frenkel, S. Ya.: Structure of macromolecules in solutions. Moscow: Nauka 1964
7. Morawetz, H.: Macromolecules in solution. New York, London: Interscience Publishers 1971
8. Flory, P. J.: Statistical mechanics of chain molecules. New York, London: Wiley Intersci. Publ. 1969
9. Yamakawa, H.: Modern theory of polymer solution. New York: Harper & Row 1971
10. Brandrup, J., Immergut, E. H. (eds.): Polymer handbook. New York: John Wiley & Sons 1975
11. Tsvetkov, V. N.: Eur. Polym. J. (Suppl.) *1969*, 237
12. Tsvetkov, V. N.: Vysokomolek. Soed. A *16*, 944 (1974)
13. Frisman, E. V., Sibileva, M. A.: Vysokomolek. Soed. *7*, 674 (1969)
14. Tsvetkov, V. N., Garmonova, T. I., Stankevitch, R. P.: Vysokomolek. Soed. *8*, 980 (1966)
15. Tsvetkov, V. N., et al.: Vysokomolek. Soed. A *13*, 884 (1971)
16. Thurston, C., Schrag, J.: J. Polym. Sci. Part A-2 *6*, 1331 (1968)
17. Burchard, W.: Makromol. Chem. *88*, 11 (1965)
18. Hearst, J. E., Stockmayer, W. H.: J. Chem. Phys. 37, 1425 (1962)

19. Hearst, J. E.: J. Chem. Phys. *38*, 1062 (1963)
20. Hearst, J. E.: J. Chem. Phys. *40*, 1506 (1964)
21. Ullmann, R.: J. Chem. Phys. *49*, 5486 (1968); *53*, 1734 (1970)
22. Burgers, J. M.: Second Report on Viscosity and Plasticity of the Amsterdam Academy of Science. New York: Nordermann Publishers 1938
23. Kirkwood, J. G., Riseman, J.: J. Chem. Phys. *16*, 565 (1948); 17, 442 (1949)
24. Tsvetkov, V. N., et al.: J. Polym. Sci. Part C *23*, 385 (1968)
25. Yamakawa, H., Stockmayer, W. H.: J. Chem. Phys. *57*, 2839 (1972)
26. Yamakawa, H., Fujii, M.: Macromolecules *6*, 407 (1973)
27. Yamakawa, H., Fujii, M.: Macromolecules *7*, 128 (1974)
28. Tsvetkov, V. N., et al.: Vysokomolek. Soed. A *10*, 547 (1968)
29. Tsvetkov, V. N., et al.: Vysokomolek. Soed. A *12*, 1892 (1970)
30. Tsvetkov, V. N., et al.: Eur. Polym. J. *7*, 1215 (1971)
31. Tsvetkov, V. N., et al.: Vysokomolek. Soed. A *14*, 369 (1972)
32. Tsvetkov, V. N.: Makromol. Chem. *160*, 1 (1972)
33. Tsvetkov, V. N., et al.: Eur. Polym. J. *9*, 27 (1973)
34. Tsvetkov, V. N., et al.: Vysokomolek. Soed. A *15*, 872 (1973)
35. Vitovskaya, M. G., et al.: Vysokomolek. Soed. B *17*, 593 (1975)
36. Andrianov, K. A., et al.: Vysokomolek. Soed. A *19*, 469 (1977)
37. Andrianov, K. A., et al.: Vysokomolek. Soed. A *20*, 1277 (1978)
38. Andreeva, L. N., et al.: Vysokomolek. Soed. *21*, 362 (1979)
39. Tsvetkov, V. N., et al.: Vysokomolek. Soed. A *16*, 566 (1974)
40. Vitovskaya, M. G., et al.: Vysokomolek. Soed. A *17*, 1917 (1975)
41. Vitovskaya, M. G., et al.: Vysokomolek. Soed. A *15*, 2549 (1973)
42. Ljubina, S. Ya., et al.: Vysokomolek. Soed. A *15*, 691 (1973)
43. Andreeva, L. N., et al.: Vysokomolek. Soed. B *17*, 326 (1975)
44. Korneeva, E. V., et al.: Vysokomolek. Soed. *21*, 1547 (1979)
45. Vitovskaya, M. G., et al.: Vysokomolek. Soed. A *18*, 691 (1976)
46. Vitovskaya, M. G., et al.: Vysokomolek. Soed. A *20*, 320 (1978)
47. Tsvetkov, V. N., et al.: Vysokomolek. Soed. *22*, 133 (1980)
48. Vitovskaya, M. G., Tsvetkov, V. N.: Eur. Polym. J. *12*, 251 (1976)
49. Tsvetkov, V. N., et al.: Dokl. Akad. Nauk. SSSR *224*, 1126 (1975); Eur. Polym. J. *12*, 517 (1976)
50. Lavrenko, P. N., Okatova, O. V.: Vysokomolek. Soed. *19*, 2640 (1977)
51. Tsvetkov, V. N., Klenin, S. I.: Dokl. Akad. Nauk SSSR *88*, 49 (1953)
52. Tsvetkov, V. N.: Ricerca Sci., Suppl. A, *25*, 413 (1955)
53. Tsvetkov, V. N., Klenin, S. I.: J. Polym. Sci. *30*, 187 (1958)
54. Vitovskaya, M. G., et al.: Vysokomolek. Soed. A *19*, 1966 (1977)
55. Vitovskaya, M. G., et al.: Vysokomolek. Soed. B *18*, 588 (1976)
56. Zimm, B.: J. Chem. Phys. *24*, 269 (1956)
57. Debye, P.: J. Chem. Phys. *14*, 636 (1946)
58. Gans, R.: Ann. Phys. (Leipzig) *86*, 628 (1928)
59. Perrin, F.: Phys. Radium *5*, 497 (1934)
60. Einstein, A.: Ann. Phys. (Leipzig) *19*, 289 (1906); *34*, 591 (1911)
61. Tsvetkov, V. N.: Vysokomolek. Soed. A *20*, 2066 (1978)
62. Kirkwood, J. G.: J. Polym. Sci. *12*, 1 (1954)
63. Riseman, J., Kirkwood, J. G. In: Rheology. Eirich, F. R. (ed.), Vol. 1, p. 495. New York: Academic Press, Interscience Publishers 1956
64. Peterlin, A., in: Rheology. Eirich, F. R. (ed.), Vol. 1, p. 615, New York: Academic Press, Interscience Publishers
65. Cerf, R.: Advan. Polym. Sci. *1*, 383 (1959)
66. Scheraga, H. A., Signer, R., in: Physical methods of organic chemistry, Weissberger, A. (ed.), Vol. 1, p. 2388. New York: Interscience Publishers 1960
67. Tsvetkov, V. N., in: Newer methods of polymer characterization. Ke, B. (ed.), p. 563. New York: Interscience Publishers 1972

68. Janeschitz-Kriegle, H.: Advan. Polym. Sci. 6, 170 (1969)
69. Peterlin, A., Munk, P., in: Physical methods of chemistry. Weissberger, A., Rossiter, B. (eds.), p. 271. New York: Interscience Publishers 1972
70. Tsvetkov, V. N., Andreeva, L. N., in: Polymer Handbook. Brandrup, J., Immergut, E. H. (eds.), 2nd Edit. New York: Wiley Intersci. Publ. 1974
71. Tsvetkov, V. N., Frisman, E. V.: Acta Physicochim. URSS 20, 61 (1945)
72. Kuhn, H.: Experienta 1, 28 (1945)
73. Kuhn, H.: Helv. Chim. Acta 31, 1677 (1948)
74. Solc, K., Stockmayer, W. H.: J. Chem. Phys. 54, 2756 (1971)
75. Solc, K.: J. Chem. Phys. 55, 335 (1971); 60, 12 (1974)
76. Mazur, I., Guttman, C., Crackin, F. Mc.: Macromolecules 6, 872 (1973); 10, 139 (1977)
77. Kranbuehl, D., Verdier, P. H., Spencer, J.: J. Chem. Phys. 59, 3861 (1973); 67, 361 (1977)
78. Linden, P. H.: J. Appl. Phys. 46, 4235 (1975)
79. Aharoni, S. H.: Polymer 19, 401 (1978)
80. Kuhn, W., Kuhn, H.: Helv. Chim. Acta 26, 1394 (1943)
81. Cerf, R.: Advan. Chem. Phys. 33, 73 (1975)
82. Tsvetkov, V. N.: Vysokomolek. Soed. 4, 894 (1962)
83. Pogodina, N. V., et al.: Vysokomolek. Soed. A 22, 2219 (1980)
84. Tsvetkov, V. N., et al.: Vysokomolek. Soed. A 10, 943 (1968)
85. Kuhn, W., Grun, G.: Kolloid Z. 101, 248 (1942)
86. Tsvetkov, V. N.: Dokl. Akad. Nauk SSSR 165, 360 (1965)
87. Hermans, J., Ullman, R.: Physica 18, 951 (1952)
88. Gotlib, Yu. Ya., Svetlov, Yu. E.: Dokl. Akad. Nauk. SSSR 168, 621 (1966)
89. Shimada, J., Yamakawa, H.: Macromolecules 9, 583 (1976)
90. Gotlib, Yu. Ya.: Vysokomolek. Soed. 6, 389 (1964)
91. Noda, I., Hearst, J. E.: J. Chem. Phys. 54, 2342 (1971)
92. Hearst, J. E., Harris, R. A.: J. Chem. Phys. 44, 2595 (1966); 45, 3106 (1966); 46, 398 (1967); 48, 537 (1968)
93. Chaffey, Ch.: J. Chim. Phys. Physicochim. Biol. 63, 1379 (1966)
94. Brown, J. F.: J. Polym. Sci. C 1, 83 (1963)
95. Andrianov, K. A., et al.: Vysokomolek. Soed. 7, 1477 (1965)
96. Andrianov, K. A., Yakushkina, S. E., Terentjeva, N. N.: Vysokomolek. Soed. A 10, 1721 (1968)
97. Andrianov, K. A.: Vysokomolek. Soed. A 11, 1362 (1969); A 13, 253 (1971)
98. Pauling, L.: Nature of the chemical bond. Ithaca, New York: Cornell University Press 1960
99. Tsvetkov, V. N., Bychkova, V. E.: Vysokomolek. Soed. 6, 600 (1964)
100. Tsvetkov, V. N., et al.: Vysokomolek. Soed. A 14, 1956 (1972)
101. Saunders, R. S.: J. Polym. Sci. A 2, 3765 (1964); A 3, 1221 (1965)
102. Nekrasov, I. K.: Vysokomolek. Soed. A 13, 1707 (1971)
103. Shashoua, V. E.: J. Amer. Chem. Soc. 81, 3156 (1959); 82, 886 (1960)
104. Schneider, N. S., Furusaki, S., Lenz, R. W.: J. Polym. Sci. A 3, 933 (1965)
105. Tsvetkov, V. N., Rjumtsev, E. I., Pogodina, N. V.: Dokl. Akad. Nauk. SSSR 224, 112 (1975)
106. Tsvetkov, V. N.: Eur. Polym. J. 12, 867 (1976); Macromolecules 11, 306 (1978)
107. Morgan, P. W.: Macromolecules 10, 1381 (1977)
108. Preston, J., in: Liquid crystalline order in polymers. Blumstein, A. (ed.). New York: Academic Press 1978
109. Andreeva, L. N., et al.: The first Meeting on Cellulose. Vladimir 1976
110. Vitovskaya, M. G., et al.: Vysokomolek. Soed. A 17, 1161 (1975)
111. Champion, J. V., Dandridge, A.: Polymer 19, 632 (1978)
112. Tsvetkov, V. N.: Dokl. Akad. Nauk, SSSR 192, 380 (1970)
113. Tsvetkov, V. N., Frisman, E. V.: Acta Physicochim. URSS 21, 978 (1946)
114. Pavlov, G. M., et al.: Vysokomolek. Soed. A 15, 1696 (1973)
115. Magarik, S. Ya., Pavlov, G. M., Fomin, G. A.: Macromolecules 11, 294 (1978)
116. Andreeva, L. N., Elokhovsky, V. Yu.: Vysokomolek. Soed. B 19, 111 (1977)
117. Schulz, G. V., Penzel, E.: Makromol. Chem. 112, 260 (1968); 113, 64 (1968)

118. Tsvetkov, V. N., Frisman, E. V., Boitsova, N. N.: Vysokomolek. Soed. *2*, 1001 (1960)
119. Tsvetkov, V. N., et al.: Vysokomolek. Soed. A *9*, 1682 (1967); A *10*, 903 (1968); A *12*, 1974 (1970); A *13*, 620 (1971); A *13*, 2532 (1971)
120. Tsvetkov, V. N., et al.: Vysokomolek. Soed. in Coll. "Cellulosa i eje Proizvodnye" *1963*, 74
121. Tsvetkov, V. N.: Usp. Khim. *38*, 1674 (1969)
122. Benoit, H.: J. Polym. Sci. *3*, 376 (1948)
123. Rjumtsev, E. I., et al.: Vysokomolek. Soed. A *17*, 2674 (1975)
124. Flory, P. J., Spurr, O. K., Karpenter, D. K.: J. Polym. Sci. *27*, 321 (1958)
125. Lavrenko, P. N., et al.: Vysokomolek. Soed. A *18*, 2579 (1976)
126. Nordermeer, J. W., Daryanani, R., Janeschitz-Kriegl, H.: Polymer *16*, 359 (1975)
127. Tsvetkov, V. N., Zakharova, E. N., Krunchak, M. M.: Vysokomolek. Soed. A *10*, 685 (1968)
128. Tsvetkov, V. N., Grishchenko, A. E., Slavetskaya, P. A.: Vysokomolek. Soed. *6*, 856 (1964)
129. Tsvetkov, V. N., Shtennikova, I. N.: Vysokomolek. Soed. *6*, 1041 (1964)
130. Sadron, C. J.: Phys. Radium *9*, 381, 384 (1938)
131. Goldstein, M.: J. Chem. Phys. *20*, 677 (1952)
132. Peterlin, A.: J. Chem. Phys. *39*, 224 (1963)
133. Daum, U.: J. Polym. Sci. A-2 *6*, 141 (1968)
134. Schulz, G. V.: Z. Phys. Chem. B *43*, 25 (1939)
135. Zimm, B. H.: J. Chem. Phys. *16*, 1099 (1948)
136. Wesslau, H.: Makromol. Chem. *20*, 111 (1956)
137. Shtennikova, I. N., Peker, T. V., Getmanchuk, Yu. P.: Vysokomolek. Soed. A *16*, 1086 (1974)
138. Shtennikova, I. N., et al.: Vysokomolek. Soed. A *20*, 1246 (1978)
139. Tsvetkov, V. N., et al.: Vysokomolek. Soed. *7*, 1104 (1965)
140. Tsvetkov, V. N., et al.: Vysokomolek. Soed. *7*, 1098 (1965)
141. Schaefgen, J. R., et al.: Polym. Prepr. *17*, 69 (1976)
142. Tsvetkov, V. N., et al.: Eur. Polym. J. *13*, 455 (1977)
143. Lavrenko, P. N., Okatova, O. V.: Vysokomolek. Soed. A *21*, 372 (1979)
144. Shtennikova, I. N., et al.: Vysokomolek. Soed. A *22*, (1980) (in press)
145. Tsvetkov, V. N., et al.: Eur. Polym. J. *14*, 475 (1978)
146. Tsvetkov, V. N., Zakharova, E. N., Mikhailova, N. A.: Dokl. Akad. Nauk SSSR *224*, 1365 (1975)
147. Tsvetkov, V. N., et al.: Vysokomolek. Soed. A *10*, 2132 (1968)
148. Tsvetkov, V. N., et al.: Eur. Polym. J. *7*, 767 (1971)
149. Tsvetkov, V. N.: Vysokomolek. Soed. A *19*, 2171 (1977)
150. Tsvetkov, V. N., Shtennikova, I. N.: Macromolecules *11*, 306 (1978)
151. Laupretre, F., Monnerie, L.: Eur. Polym. J. *14*, 415 (1978)
152. Tsvetkov, V. N., et al.: Vysokomolek. Soed. A *21*, 1711 (1979)
153. Koton, M. M.: Vysokomolek. Soed. A *16*, 1199 (1974)
154. Tsvetkov, V. N., Pogodina, N. V., Malichenko, B. F.: Vysokomolek. Soed. A *21*, 564 (1979)
155. Tsvetkov, V. N., et al.: Acta Polymerica *31*, 434 (1980)
156. Tsvetkov, V. N., et al.: Vysokomolek. Soed. A *21*, 83 (1979); Macromolecules *12*, 645 (1979)
157. Tsvetkov, V. N., Pogodina, N. V., Starchenko, L. V.: Eur. Polym. J. *16*, 387 (1980)
158. Tsvetkov, V. N., et al.: Vysokomolek. Soed. A *11*, 349 (1969)
159. Tsvetkov, V. N., et al.: Dokl. Akad. Nauk. SSSR *205*, 895 (1972)
160. Andreeva, L. N., et al.: Vysokomolek. Soed. B *15*, 209 (1973)
161. Philippoff, W., Tornqvist, E. G. M.: J. Polym. Sci. C *23*, 881 (1969)
162. Tsvetkov, V. N., et al.: Vysokomolek. Soed. A *11*, 2528 (1969)
163. Tsvetkov, V. N., et al.: Eur. Polym. J. *9*, 481 (1973)
164. Tsvetkov, V. N., et al.: Vysokomolek. Soed. *5*, 3 (1963)
165. Tsvetkov, V. N., et al.: Vysokomolek. Soed. A *13*, 2011 (1971)
166. Tsvetkov, V. N., Marinin, V. A.: J. Exp. Theor. Phys. (SSSR) *18*, 641 (1948)
167. O'Konski, C. T. (ed.): Molecular electro-optics. New York, Basel: Marcel Dekker, Interscience Publishers 1976

168. Marinin, V. A., Poljakova, L. V., Korolkova, Z. S.: Univ. Vestn. Ser. Fiz. Khim. (Leningrad) *16*, 73 (1958)
169. Champion, J. V., Meeten, G. H., Southwell, G. W.: Polymer *17*, 651 (1976)
170. Tsvetkov, V. N., Petrova, A. I.: Zh. Tekhn. Fiz. *14*, 289 (1944)
171. Le-Fevre, C. G., Le-Fevre, R. I. W., Parkins, G. M.: J. Chem. Soc. *1958*, 1468; *1960*, 1814; *1960*, 2890
172. Le-Fevre, R. I. W., Saundaram, K. M. S.: J. Chem. Soc. *1963*, 1880, 3188, 3547
173. Dous, D.: J. Chem. Phys. *41*, 2656 (1964)
174. Nagai, K., Ishikawa, I.: J. Chem. Phys. *43*, 4508 (1965)
175. Le-Fevre, R. I. W., Sundaram, K. M. S.: J. Chem. Soc. *1962*, 1494
176. Le-Fevre, R. I. W., Sundaram, K. M. S.: J. Chem. Soc. *1962*, 4003
177. Doty, P., Wagner, H., Singer, S.: J. Phys. Colloid Chem. *51*, 32 (1947)
178. Arlman, E. Y., Boog, W., Coumon, D. J.: J. Polym. Sci. *16*, 543 (1953)
179. Hengstenberg, J., Schuch, E.: Makromol. Chem. *74*, 55 (1964)
180. Stuart, H. A., Peterlin, A.: J. Polym. Sci. *5*, 551 (1950)
181. Langevin, P.: Compt. Rend. *151*, 475 (1910); Radium *7*, 249 (1910)
182. Born, M.: Ann. Phys. (Leipzig) *55*, 177 (1918)
183. Volkenstein, M. V.: Configurational statistics of polymeric chains. New York: Interscience Publishers 1963
184. Kelly, K. M., Patterson, G. D., Tonelli, A. E.: Macromolecules *10*, 859 (1977)
185. O'Konski, C. T., Yoshioka, K., Orttung, W. H.: J. Phys. Chem. *63*, 1558 (1959)
186. Nakayama, H., Yoshioka, K.: J. Polym. Sci. A *3*, 813 (1965)
187. Yoshioka, K., O'Konski, C. T.: J. Polym. Sci. A-2 *6*, 421 (1968)
188. Kikuchi, K., Yoshioka, K.: J. Phys. Chem. *77*, 2101 (1973)
189. Tsvetkov, V. N., Ljubina, S. Ya., Bolevsky, K. L.: Vysokomolek. Soed., in Coll. "Carbotseprnye Polymeri" 33, 1963
190. Tsvetkov, V. N., Ljubina, S. Ya., Barskaja, T. V.: Vysokomolek. Soed. *6*, 806 (1964)
191. Tsvetkov, V. N., et al.: Vysokomolek. Soed. *8*, 846 (1966)
192. Tricot, M., Houssier, C., Desreux, V.: Eur. Polym. J. *12*, 575 (1976); *14*, 307 (1978)
193. Foweraker, A. R., Jennings, B. R.: Polymer *16*, 720 (1975)
194. Tricot, M., Houssier, C., Desreux, V.: Biophys. Chem. *8*, 221 (1978)
195. Benoit, H.: Anal. Phys. (Paris) *6*, 561 (1951)
196. Hanss, M., Bernengo, J.: Biopolymers *12*, 2151 (1973)
197. Miller, S., Wetmur, J.: Biopolymers *13*, 115 (1974)
198. Matsumoto, M., Watanabe, H., Yoshioka, K.: Biopolymers *6*, 929 (1968); *12*, 1729 (1973)
199. Kikuchi, K., Yochioka, K.: Biopolymers *12*, 2667 (1973)
200. Tsvetkov, V. N., et al.: Vysokomolek. Soed. *5*, 453 (1963)
201. Watanabe, H., Yoshioka, K., Wada, A.: Biopolymers *2*, 91 (1964)
202. Tsvetkov, V. N., et al.: Vysokomolek. Soed. *7*, 1111 (1965)
203. Tsvetkov, V. N., et al.: Vysokomolek. Soed. *8*, 1466 (1966)
204. Tsvetkov, V. N., et al.: Vysokomolek. Soed. A *9*, 1575, 1583 (1967)
205. Tsvetkov, V. N., et al.: J. Polym. Sci. C *16*, 3205 (1968)
206. Ohe, H., Watanabe, H., Yoshioka, K.: Colloid and Polym. Sci. Darmstadt *252*, 26 (1974)
207. Nishioka, M., Kikuchi, K., Yoshioka, K.: Polymer *16*, 791 (1975)
208. Boeckel, G., et al.: J. Chim. Phys., Phys. Chim. Biol. *59*, 999 (1962)
209. Peterlin, A., Stuart, H. A.: Hand- und Jahrbuch der chemischen Physik vol. 8, part 1B, p. 24. Leipzig: Akademische Verlagsgesellschaft, Geest & Portig
210. Tsvetkov, V. N., et al.: Eur. Polym. J. *11*, 37 (1975)
211. Rjumtsev, E. I., Pogodina, N. V., Getmanchuk, Yu. P.: Vysokomolek. Soed. A *17*, 1719 (1975)
212. Rjumtsev, E. I., Aliev, F. M., Tsvetkov, V. N.: Vysokomolek. Soed. A *17*, 1712 (1975)
213. Tsvetkov, V. N., et al.: Eur. Polym. J. *10*, 563 (1974)
214. Lavrenko, P. N., et al.: J. Polym. Sci. Symp. *44*, 217 (1974)
215. Pogodina, N. V., et al.: Vysokomolek. Soed. B *19*, 851 (1976)
216. Tsvetkov, V. N., et al.: Vysokomolek. Soed. A *17*, 2493 (1975)
217. Tsvetkov, V. N., et al.: Eur. Polym. J. *7*, 767 (1971)

218. Tsvetkov, V. N., et al.: Eur. Polym. J. *9*, 1 (1973)
219. Tsvetkov, V. N., et al.: Vysokomolek. Soed. A *15*, 400 (1973)
220. Stockmayer, W. H., Baur, M. E.: J. Amer. Chem. Soc. *86*, 3485 (1964)
221. Stockmayer, W. H.: Pure Appl. Chem. *15*, 539 (1967)
222. Stockmayer, W. H.: Macromolecules *2*, 647 (1969); *5*, 766 (1972)
223. Rjumtsev, E. I., et al.: Vysokomolek. Soed. A *17*, 1368 (1975)
224. Zimm, B., Roe, G., Epstein, L.: J. Chem. Phys. *24*, 279 (1956)
225. Wilkinson, R. S., Thurston, G. B.: Biopolymers *15*, 1555 (1976)
226. Kuhn, W., Dührkop, N., Martin, H.: Z. Phys. Chem. *45*, 121 (1939)
227. Rjumtsev, E. I., et al.: Vysokomolek. Soed. A *17*, 61 (1975)
228. Tsvetkov, V. N., Rjumtsev, E. I., Pogodina, N. V.: Vysokomolek. Soed. A *19*, 2141 (1977)
229. Tsvetkov, V. N., et al.: Eur. Polym. J. *10*, 55 (1974)
230. Kuhn, W.: Helv. Chim. Acta *31*, 1092 (1948)
231. Tsvetkov, V. N.: Dokl. Akad. Nauk. SSSR *205*, 328 (1972)
232. Rjumtsev, E. I., et al.: Vysokomolek. Soed. A *18*, 439 (1976)
233. Tsvetkov, V. N., et al.: Vysokomolek. Soed. A *18*, 2016 (1976)
234. Tsvetkov, V. N., et al.: Vysokomolek. Soed. A *15*, 2270 (1973)
235. Mukohata, Y., Ikeda, S., Isemura, T.: J. Mol. Biol. *5*, 570 (1973)
236. Ikeda, S.: J. Chem. Phys. *38*, 1062 (1963)
237. Mukohata, Y.: J. Mol. Biol. *7*, 442 (1963)
238. Tsvetkov, V. N., Vinogradov, E. L.: Opt. Spectrosk. *21*, 603 (1966)
239. Tsvetkov, V. N., Vinogradov, E. L.: Vysokomolek. Soed. *8*, 662 (1966)
240. Le-Fevre, C., Le-Fevre, R., in: Technique of organic chemistry. Weissberger, A. (ed.), Vol. 1, part 3, chapter 36. New York: Interscience Publishers 1960
241. Frederiq, E., Houssier, C.: Electric Dichroism and Electric Birefringence. Oxford: Clarendon 1973
242. Taylor, G.: Proc. Roy. Soc. A *157*, 546, 565 (1936)
243. Frisman, E. V., Tsvetkov, V. N.: J. Exp. Theor. Phys. (SSSR) *23*, 690 (1952)
244. Frisman, E. V., Tsvetkov, V. N.: Zh. Tekhn. Fiz. (SSSR) *25*, 447 (1955)
245. Frisman,, E. V., Sjui-Mao: Vysokomolek. Soed. *3*, 276 (1961)
246. Brace, D. B.: Phys. Rev. *18*, 70 (1904); *19*, 218 (1904)
247. Szivessy, G., in: Handbuch der Physik. Geiger, H., Scheel, K. (eds.), Vol. 19, p. 918. Berlin: Springer Verlag 1928
248. Pen'kov, S. N., Stepanenko, B. Z.: Opt. Spectrosk. *14*, 156 (1963)
249. Tsvetkov, V. N.: J. Exp. Theor. Phys. *21*, 701 (1951)
250. Tsvetkov, V. N.: Vysokomolek. Soed. A *19*, 1249 (1967)
251. Tsvetkov, V. N., Skazka, V. S., Lavrenko, P. N.: Vysokomolek. Soed. A *13*, 2251 (1971)
252. Gray, G. W.: Molecular structure and properties of liquid crystals. New York: Academic Press 1962.
253. Robinson, C.: Trans. Faraday Soc. *52*, 571 (1956)
254. Samulski, E. T., in: Liquid crystalline order in polymers. Blumstein, A. (ed.), p. 167. New York: Academic Press 1978
255. Aharoni, S. M.: Macromolecules *12*, 94 (1979)
256. Aharoni, S. M., Walsh, E. K.: Macromolecules *12*, 271 (1979); J. Polym. Sci., Polym. Lett. *17*, 321 (1979)

Received March 12, 1980
K. Dusek (editor)

Author Index Volumes 1–39

Polymers

Properties and Applications

Editorial Board:
H. J. Cantow, H. J. Harwood,
J. P. Kennedy, A. Ledwith,
J. Meißner, S. Okamura,
G. Olivé, S. Olivé

Springer-Verlag
Berlin
Heidelberg
New York

Volume 1: B. Rånby, J. F. Rabek

ESR Spectroscopy in Polymer Research

1977. 356 figures, 29 tables. XIV, 410 pages
ISBN 3-540-08151-8

"...This book is a remarkable example for the successful combination of simplicity and clarity in its tutorial parts and of depth and width whenever and wherever it presents the state of the art... As ultimate and very gratifying reward for his investment the reader gets no less than 2519 references to the literature in excellent alphabetical order. Scientists who already work with ESR will be greatly assisted in their efforts by this book; those who do not yet use this method will have an easy time to learn and use it. All of them will be grateful to the authors for this exceptional addition to our scientific literature."
J. Polymer Science

Volume 2: H.-H. Kausch

Polymer Fracture

1978. 180 figures, 23 tables. X, 332 pages
ISBN 3-540-08786-9

"...The avowed aim of this book is to connect the more conventional statistical and continuum mechanics interpretation of fracture phenomena to the newer spectroscopic studies of highly stressed polymeric chains and the kinetics of their rupture. Relating the literature on the observed modes of viscoelasticity and irreversible deformation from polymer morphology and solid-state physics. Kausch explains the behavior and rupture of polymeric materials in terms of molecular slip and breakage processes. This leads to interesting, methodical and well-thought-out interpretations of fracture toughness, crack propagation rates and fatigue of all major polymer systems... Thus, the book is an outstanding contribution to our understanding of the role of chain ruptures during mechanical failure..."
Physics Today

Volume 3: A. Knop, W. Scheib

Chemistry and Application of Phenolic Resins

1979. 111 figures, 88 tables. XIII, 269 pages
ISBN 3-540-09051-7

The authors present the current theory of phenolic resin chemistry and the technical application of phenolic resins, based an day-to-day experience in research, production and marketing, and against the background of economic relevance. Where the first fully synthetic polymers (phenolic resins) stand today and what their future is are subjects of discussion. Looking back at their development, it is shown that after a wide variety of adaptions, they remain technically and economically irreplaceable products with potential for further market growth and a commensurate appreciation of their value.

Volume 4: A. Hebeish, J. T. Guthrie

The Chemistry and Technology of Cellulosic Copolymers

1980. 91 figures, approx. 91 tables. Approx. 500 pages
ISBN 3-540-10164-0

This monograph provides an informative account of new, improved cellulosic materials and the chemistry and technology involved in their production, as well as the first detailed description of grafted and modified celluloses.
The information contained in this book will be of great value to researchers, manufacturers, but also instructors, interested in the modification of cellulosics for textiles, paper, printing, printing inks, paints, and packaging, as well as in polymerization processes and cellulose derivativization.

Polymer Bulletin

Editors:
Prof. H.-J. Cantow, Makromolekulare Chemie, Universität Freiburg, Stefan-Meier-Strasse 31,
D-7800 Freiburg, West-Germany
Prof. J. P. Kennedy, Dept. of Polymer Science, The University of Akron, Akron, OH 44325, USA
Prof. T. Saegusa, Dept. Synthetic Chemistry, Kyoto University, Kyoto, 606, Japan

Editorial Board: H. Batzer, Basel; N. Calderon, Akron, OH; S. Cesca, San Donato Milanese; P. J. Flory, Stanford, CA; J. Furukawa, Tokyo; J. E. McGrath, Blacksburg, VA; H. K. Hall, Jr., Tucson, AZ; H. H. Kausch, Lausanne; T. Kelen, Budapest; M. Kryszewski, Lódź; A. Ledwith, Liverpool; E. Maréchal, Paris-Cedex; J. Meißner, Zürich; A. Nakajima, Kyoto; G. and S. Henrici Olivé, Research Triangle Park, NC; N. A. Plate, Moscow; B. Rånby, Stockholm; C. I. Simionescu, Bucureşti; S. Sivaram, Gujarat; D. H. Solomon, Melbourne; R. Steiner, Frankfurt/M.; H. Tadokoro, Osaka; M. Takayanagi, Fukuoka; I. Uematsu, Tokyo; C. Wippler, Strasbourg; H. Zahn, Aachen

Editorial Assistant: A. Heinrich, Springer-Verlag Heidelberg

To cope with the rapid progress of polymer science, a new journal is now published characterized by emphasis on rapid publication of papers containing a most concise description of results.
The character of the new journal is between the purely archival journals of full papers and the so-called "letter journals" consisting exclusively of short communications.

Special features:
- rapid publication of papers
- no page charge
- 50 off-prints of each paper supplied free of charge

Subscription information and sample copy upon request

Send your order to your bookseller or directly to: Springer-Verlag, Journal Promotion Dept., P. O. Box 105280, D-6900 Heidelberg, FRG

North America: Springer-Verlag New York Inc., Journal Sales Dept., 44 Hartz Way, Secaucus, NJ 07094, USA

Springer
International